The Phenomenology of Traffic

The book delves into the affective, embodied, and sensory dimensions of traffic and urban mobility. It brings together key phenomenological and post-phenomenological readings to challenge taken-for-granted assumptions of urban traffic.

Through the experiences of traffic users in Ho Chi Minh City, Vietnam, the book provides fascinating pathways into structures and processes that make up phenomenal traffic worlds. It explores the nature of the traffic experience, modalities of existence within it, and the wide spectrum of awarenesses involved in making sense from non-sense. The book offers rich theoretical insights on how we feel our way through our affect-laden worlds. Through empirical examples from the urban traffic in Ho Chi Minh City, the book explores this fluid, constantly changing complex collective of ongoing negotiations we call 'traffic,' often emotional, involving and producing all kinds of entities. It develops a range of philosophical concepts in order to better understand the complex relationships between humans and nonhumans in everyday settings.

Offering innovative insights into the structures, authorities, materialities, and forms of power that shape our experiences of traffic, this book will be of interest to students, scholars, and practitioners interested in philosophy, cultural geography, mobilities, transport studies, cultural studies, and urban studies.

Glenn Wyatt is a designer, illustrator, and lecturer with an interest in books, art, music, philosophy, and affect. He has lived in many countries, including Vietnam, Morocco, Italy, and Thailand.

Ambiances, Atmospheres and Sensory Experiences of Spaces

Series Editors:
Rainer Kazig, CNRS Research Laboratory Ambiances – Architectures –
Urbanités, Grenoble, France
Damien Masson, Université de Cergy-Pontoise, France
Paul Simpson, Plymouth University, UK

Research on ambiances and atmospheres has grown significantly in recent years in a range of disciplines, including Francophone architecture and urban studies, German research related to philosophy and aesthetics, and a growing range of Anglophone research on affective atmospheres within human geography and sociology.

This series offers a forum for research that engages with questions around ambiances and atmospheres in exploring their significances in understanding social life. Each book in the series advances some combination of theoretical understandings, practical knowledges and methodological approaches. More specifically, a range of key questions which contributions to the series seek to address includes:

- In what ways do ambiances and atmospheres play a part in the unfolding of social life in a variety of settings?
- What kinds of ethical, aesthetic, and political possibilities might be opened up and cultivated through a focus on atmospheres/ambiances?
- How do actors such as planners, architects, managers, commercial interests and public authorities actively engage with ambiances and atmospheres or seek to shape them? How might these ambiances and atmospheres be reshaped towards critical ends?
- What original forms of representations can be found today to (re)present the sensory, the atmospheric, the experiential? What sort of writing, modes of expression, or vocabulary is required? What research methodologies and practices might we employ in engaging with ambiances and atmospheres?

The Phenomenology of Traffic
Experiencing Mobility in Ho Chi Minh City
Glenn Wyatt

For more information about this series, please visit: https://www.routledge.com

The Phenomenology of Traffic
Experiencing Mobility in Ho Chi Minh City

Glenn Wyatt

Routledge
Taylor & Francis Group

LONDON AND NEW YORK

First published 2021
by Routledge
2 Park Square, Milton Park, Abingdon, Oxon OX14 4RN

and by Routledge
605 Third Avenue, New York, NY 10017

First issued in paperback 2022

Routledge is an imprint of the Taylor & Francis Group, an informa business

Publisher's Note
The publisher has gone to great lengths to ensure the quality of this
reprint but points out that some imperfections in the original copies may
be apparent.

British Library Cataloguing-in-Publication Data
A catalogue record for this book is available from the British Library

Library of Congress Cataloging-in-Publication Data
A catalog record has been requested for this book

ISBN 13: 978-0-367-54655-7 (pbk)
ISBN 13: 978-0-367-21816-4 (hbk)
ISBN 13: 978-0-429-26628-7 (ebk)

DOI: 10.4324/9780429266287

Typeset in Times New Roman
by codeMantra

Contents

Figures

Acknowledgements

Some months ago, a friend and fellow expatriate in Ho Chi Minh City (HCMC) shared the news of his imminent plans to finally leave the city. His many years of accumulated experience here culminated in the somewhat resolute and definitive statement, something to the effect of, everything seems normal, but then you realize it's really not. Living in Vietnam as an expatriate is not for everyone; indeed, it may not even be for me, and yet I have stayed on for more than 10 years, as life continues to oscillate between the kinds of polar extremes that are seemingly part and parcel of living in a place that is both passionate and does not negotiate. Life in Vietnam can teach one many things, not least of all courage and resilience, two characteristics that Vietnamese people tend to admire, and foreigners who live in Vietnam generally fall into three different groups: those who leave after only a year or two, those who leave after 10 or 20 years, and those who never leave.

Though it exists under a single name and as one unified country, Vietnam is really a mix of quite different cultural areas, such as between the north and the south, and it is in the complex melting pot of HCMC that all of these mindsets, practices, tendencies, and attitudes intermingle and co-exist. My experience in Vietnam has been almost exclusively that of the south and of HCMC. When I first moved here, on perhaps only my second day, my colleague took me around the city on her motorbike. I remember being struck by an atmosphere so energetic and full of life that it seemed like a party, and yet it was neither a special event, nor even a Friday or Saturday night. There seemed just so much to take in with every glance that I didn't know where to begin; it was all movement and sound, so overwhelming that I found myself just treading water, as from looking down from a great height, whilst fighting the urge to jump. Like many others, what struck me was the traffic, which was then, aside from taxis, almost entirely made up of motorbikes; I remember saying to my new colleague, "I bet one can learn a lot about the Vietnamese just by looking at the traffic."

Since that day, I have racked up tens of thousands of kilometres on the only motorbike I have ever owned, my trusty red 125cc Honda Airblade. I have worked as a motorbike courier for my own enterprises, transporting

x   *Acknowledgements*

great loads of wood and other materials across all districts of the city from suppliers to factories, all strapped to the side of the motorbike, the weight of which, on one occasion, almost toppled me in front of an oncoming bus. Though I have witnessed countless numbers of traffic accidents, both minor and not so minor, miraculously, I have never had an accident that resulted in injury, but the near misses are too numerous to mention, and the probabilities are increasingly stacked in the wrong ledger column.

HCMC is filled with invitations and celebrations. It is filled with passion, love, hate, empathy, cold-heartedness and genuine warmth, and the constant buzz of sociality and of humanity on the move. It is a city of open doors that remain stubbornly closed, smiles and families, the sounds of children playing, and the noise of building construction, of freedom and football. It is a city of dreams and opportunities, and, yes, traffic. Journeys in Vietnam, into traffic or otherwise, require guides, and I am very grateful to those who have shown me things that, though they were always right in front of my eyes, I would never had known. I am grateful for all those who shared their experiences in traffic with me and especially for the assistance of Hien Nguyen and Hieu Hue Lieu for helping me to traverse the cultural complexities involved in research in Vietnam. I am also greatly indebted to Nguyet Huynh for her ongoing support, patience, and tireless efforts in dotting 'i's and crossing 't's. In a city constituted of constant tectonic shifts and change, I am grateful for the support, stability, and loyalty of Chau Tran, without whom this book would surely not exist. Whether eating goat hotpot by the side of the road or sampling local beer, together, like twin bookends, like two sentinel statues, such as often seen at the entrances to buildings, we look on, never missing a trick and always honing our senses to the winds of change that continue to blow through this amazing city.

I would like to thank Faye Leerink at Routledge for her initial interest in the idea of this book and Nonita Saha for answering all my editorial-related questions. I am especially grateful for the constantly surprising level of support from Professor Mandy Thomas (who has her own traffic stories of Vietnam to tell), who somehow finds the time in her busy schedule as Executive Dean to read and comment on the thousands of words I keep sending to her. I would also like to thank Keith Robinson, who gave me timely critical feedback on the elements of process philosophy I was then (and still) wrangling with, and Catherine Earl, whose critical comments on parts of my final draft inspired necessary last-minute changes. Finally, for my family, who are all great travellers, no matter how far I roam, they still manage to appear at my doorstep for a visit. Though we rarely seem to share the same physical space, they are, nevertheless, always with me, and throughout the journey of this book and all other journeys, I am most grateful for their unfailing support and guidance along the road.

Preface

As a subject of investigation, the urban traffic in Ho Chi Minh City (HCMC) gives valuable insight into the nature of change and of how complex things are resolved and continually reinforced, thereby gaining stability over time. In this way, enduring things, as structures and patterns or as faded ghosts in varying guises, influence future outcomes and come to be an experience *of* time. In this book, the HCMC traffic, with the help of an integrated theoretical framework, provides fascinating pathways into structures and processes that make up phenomenal traffic worlds. Therefore, this is not really a book about traffic *per se*, but about the nature of experience itself and the wide spectrum of awarenesses involved in making sense from non-sense, as we feel our way through our affect-laden worlds. The choice of the HCMC traffic as the subject of these investigations is far from incidental, for this traffic system is a fluid, constantly changing complex collective of ongoing negotiations, often emotional, involving and producing all kinds of entities. With the right lens, the ghostly presences of traffic – immaterial flows, atmospheres, and affective timespaces – come to be foregrounded, revealing how the immaterial flows through the material and vice versa, coagulating, coalescing, and inspiring new novel events. The HCMC traffic is life in action, where what we call the 'social' and the 'technological' merge with and challenge our assumptions of that which we call the 'natural' realm.

After more than 10 years of experience living with, observing, and participating in this traffic system, I still find it constantly surprising. As with many experiences when living as a foreign expatriate in Vietnam, just when you feel confident to bask in the warm waters of understanding and clarity, the waters muddy again and you realize how little you know, how much there is to know, and how much can never be known. This book is the culmination of more than four years of empirical and academic research and more than a year spent writing and researching the book. It is clearly apparent that the traffic system that now exists in HCMC differs enormously from the one I first began researching, and even more profoundly from the traffic I first experienced here 10 years ago. Where once there were motorcycles dominating the landscape, there are now many more cars. Where once there were freelance motorcycle taxi riders sitting on street corners waiting patiently for customers, there are now networks and infrastructures of Internet-based ride hailing applications, such as Grab, efficient and technologically complex.

Vietnam has a complex history, to say the least, and these past events continue to resonate, influencing certain kinds of outcomes in the present and beyond. Contemporary Vietnam, and, of note here, its urban traffic, is changing rapidly and exponentially through myriads of novel divergences, streaming towards myriads of futures, as the past inspires new becomings. As Grosz (2004) says, life always brings the resources of the past, of habit, instinct, memory, and learning to life's movement towards differentiation, which is "always directed to elaboration, complication, emergence," and to the "forward pull of time" (p. 255). The past, then, provides not only the conditions for the present but also the conditions for all possible, increasingly complex, futures that arise from the present, not as cause, but as virtuality and potential (Grosz, 2004); in other words, the past inspires, but in nonlinear ways that could never be predicted.

Novelty, the nature of experience, awareness, perception, and the experience of the passing of time, those processual *blackboxes*, where infinite potentiality gets knitted together into concretely existing actuality, are central concerns for phenomenology. The following comment from a traffic user in HCMC poetically illustrates the experience of riding a motorcycle in the HCMC traffic. This motorcyclist, who was new to HCMC and had only been riding in the city for about two months, commented to me some years ago, "You need to have five eyes to look around." At first glance, the traffic of HCMC appears extremely chaotic and disordered, but with experience, this apparent chaos begins to gain structure and order. Of course, a glance is a different mode of awareness to that of *being-in* and being emotionally involved, with something at stake. Successful participation and survival in this traffic collective requires the involvement of a wide spectrum of awarenesses. Events, and what Levinas calls "the staging of that which is the object" (as cited in Moran, 2000, p. 327), are filled with affect, ambience, and atmosphere, and the involvement of these more primordial pathways of negotiation, decision-making, and experience in traffic should not be underestimated. Non-representational dimensions such as emotions, tacit knowledge, the realm of the senses, and those vague feelings of the pre-conscious are fundamental to how we feel our way through traffic, as opposed to how we feel *about* traffic. The HCMC traffic system, with its history of bicycles and motorcycles, has evolved an aesthetic style fundamentally influenced by the movements and affordances of these kinds of vehicles as well as deep ways of being and cultural beliefs and practices. Its structure includes sets of constantly changing tacit collective agreements and informal rules and practices that exist alongside the formal traffic laws. Therefore, attunement to the ebb and flow of this system requires the development of awarenesses that are sensitive to subtle ways of being, affect, and atmosphere.

The structure of this book is such that each chapter builds upon the previous, more like a novel, gradually deepening and widening the scope of the subject matter over the course of the book. Therefore, it is recommended that the reader begin at the beginning and hang in there until the end, as the concepts build upon each other and open up new horizons, marrying theory and real-life empirical stories. Whilst it is common practice in academic books to summarize each chapter at the beginning and end of the chapter in question, I have chosen not to do so, but to let things emerge as they emerge, rather than expose the whole plot before we have even begun.

However, the following section provides short descriptions of each chapter. Chapter 1 introduces the aesthetic and general character of the traffic in HCMC, initially through the experience of a Vietnamese motorcycle rider. Drawing on the process philosophy of Alfred North Whitehead, the chapter then raises the question of the relationship between the past and the creative evolution of present events within frameworks of kinetics, affect, and atmosphere. Chapter 2 outlines the aims and goals of the book. By way of an affective large-scale traffic event, this chapter describes some of the dynamics and forces that characterize the traffic in HCMC, suggesting the need for a more phenomenological, relational, and non-representational approach to the study of traffic. The chapter also outlines and contextualizes the approach and framework formulated in this book within the existing landscapes of current traffic research, for example, mobilities studies, as well as less sociological, more technological approaches, such as computer modelling and simulations. Chapter 3 focuses on the specific history of Vietnam and speculates on the ways in which past events continue to resonate through evolutionary processes that mould the traffic system in contemporary HCMC, from micro events to large-scale infrastructure projects. Chapter 4 describes Whitehead's notion of *the bifurcation of nature*, a view of the world as constituted, on one side, by universal objectifications of knowledge and scientific facts, and, on the other, as an experiential realm of affective intensities, emotions, sensorial experiences, and subjective concerns and values. The relevance of these ideas, especially in regard to the representation and experience of entities, objects, and infrastructures, is then discussed within the context of dominant traffic research approaches. Chapter 5 focuses on the notion and science of complexity and its potential for understanding the creative evolutionary processes of complex urban phenomena such as traffic systems, especially in relation to affect and atmosphere, and the project of the development of a *metaphysics of experiential complexity*. Drawing on affect theorists, such as Massumi and Manning, the embodied phenomenology of Merleau-Ponty, and the ontology of 'feeling-tones,' as described in Whitehead's process philosophy, Chapter 6 draws us deeper into the nature of objects through empirical traffic events in HCMC. Staying with objects and events, Chapter 7 builds bridges between the eternal and the actual and between the past and the future, through discussions of the substantializing power of the immaterial and the abstract. Chapter 8 investigates the notion of the 'subject' as both *feeler* and the *feelings being felt*, as an emergent outcome dwelling deep within infrastructures and networks of atmosphere and affect, and what Whitehead calls *lures for feeling*. Through empirical HCMC traffic stories, Chapter 9 further describes the ways in which HCMC traffic users navigate a phenomenon fundamentally constituted by rhythms, forms, intensities, and resonances, integrated into networks of nonhumans, using kinaesthetic sensibilities. Drawing on the notion of autopoiesis and Whitehead's processes of *concrescence* and *transmutation*, Chapter 10, through empirical examples from HCMC traffic, explores how the infinite possibilities of random potential come to be experienced synaesthetically and kinaesthetically *as order*, and as relationalities imbued with meaning and significance, all manifested in a cohesive aesthetic style.

References

Grosz, E. (2004). *The nick of time: Politics, evolution, and the untimely*. Durham, NC: Duke University Press.

Moran, D. (2000). *Introduction to phenomenology*. Abingdon, England: Routledge.

1 The experience of traffic in Ho Chi Minh City

Recently, I met with a Vietnamese friend after work, in a quiet bar, conveniently located to both our apartments. Like most other traffic users in Ho Chi Minh City (HCMC), to get from home to work and vice versa, she rides her motorcycle through traffic conditions, the likes of which citizens in many other cities could not even imagine. This daily practice that one might refer to as the 'mundane commute,' in HCMC, comes with no small amount of danger and accumulated ill effects on health. Riding a motorcycle in HCMC might be a mundane activity, but that doesn't mean that it is not a dangerous (and, perhaps, slightly crazy) thing to do, given the risks. For most, driving a car, or taking a taxi or Grab car is not economically viable, and given the lack of public transportation services in the city, my friend, like most others (like myself), has little choice but to enter this typhoon of sociotechnical manic activity, every day, and generally without complaint.

Over the many years we have known each other, my friend has, on occasion, shared with me some nostalgic memories of her childhood, of growing up in HCMC, where just to pass the time, she often leisurely rode her bicycle great distances around the city. Those nostalgic days of riding for leisure now seem firmly in the past. In HCMC, participation in the traffic forms a substantial and challenging dimension to one's existence, and every moment my friend spends in the traffic, she contributes, unknowingly, to its current form, as it, equally unknowingly, contributes to the person I know as Châu. In the traffic in HCMC, we all "find ourselves in a buzzing world, amid a democracy of fellow creatures" (Whitehead, 1978, p. 50), in a world full of actualities that exist in the same sense as we do (Hosinski, 1993), vying for the right to move forward, but in a democracy where physical size matters.

From inside the bar, I watch as she parks her scooter on the footpath and removes some of her traffic apparel: helmet, gloves, 'hoodie' jacket, and facemask. She wears less additional 'traffic' clothing than many other female motorcyclists in the city, who have earned the name 'ninjas,' wrapped as they often are, from head-to-toe in traffic accessories that are worn over their regular clothes; stealthful and anonymous, yet in less clandestine

Figure 1.1 Negotiating for mobility rights in a common space. Copyright 2016 by
       G. Wyatt.

ways, somehow always managing to get in my way. As Châu enters the air-
conditioned bar, with its warm yellow-lit ambience and quiet music, the
traffic noise from outside briefly enters with her through the open door,
temporarily changing the mood inside, a reminder of the vast difference
between the two worlds. I turn to greet her, and, when closer, I am struck by
her expression. I see signs of accumulated stress and exhaustion, so much so,
in fact, that her face is almost frozen into a mask, as though she has just experienced a haunting. I know this person to be extraordinarily patient and
forgiving, but I sense an imminent implosion or explosion, and so scurry to
find her a glass of wine.

   She sits down next to me, still looking somewhat harassed as her eyes stare
out, unfocused, and unseeing. As she stares blankly, her body is still, but I
imagine her mind is anything but immobile, perhaps still buzzing with ghost
motorcycles, the constant beeping of horns, the smell of exhaust fumes, the
roar of engines, and the kamikaze buses that career through the HCMC
streets, forging their own paths like ice-breaker ships or snow ploughs. She
removes her jacket slowly and consciously, as though intentionally *excom-
muning*, as though it were a sacred rite that might shed the layers of dust,
heat, and noise from which she has just been extricated. Yet, the quiet ambience of the bar still remains beyond her; she is yet to find it, yet to dwell in it.
As her eyes and brain begin to focus again on the here and now, she picks up
a menu from the table, but before considering its contents, turns to me, and
with some resignation, but still that defiant resilience that comes with being
Vietnamese, she says, "if Vietnam ever has another war, I know we will win
again." In a culture that prizes courage and confidence and does not shrink
from a fight, vying for mobility rights can be a serious affair.

Edensor (2010) notes that "place is characterized by the mobilities that course through it" (p. 5), and the aesthetic character of HCMC would, no doubt, be a very different kind of place without its particular version of a traffic system.

For people new to HCMC (as well as some who are not so new), this is traffic that challenges our conceptions of what 'traffic' is or even *should be*, and it can be a common topic of conversation amongst Vietnamese locals and expatriates alike. Visitors to the city are often seen standing by the roadside taking videos of the traffic, as though they were witnessing the migration herds of wild animals, or simply looking perplexed, caught like an insect in an amber bubble of indecision and confusion, simply unable to figure out just how to cross the street. I suspect that this fascination resides in the fact that while the traffic contains the usual universal paraphernalia of traffic: traffic lights, stop signs, cars, etc., it operates according to a logic different to traffic in other cities. In other words, whilst it still looks like a duck, it doesn't really seem to walk like one, and oftentimes, I hear visitors make comment that they are surprised not to have seen an accident or that "it seems to work," a sentence uttered in complete amazement.

Mobility in the HCMC traffic system can be less like *commuting*, and more like surviving, striving, being prepared to leave everything on the fighting room floor if that is what it takes. This sounds negative, but, in many ways, the experience of being in HCMC traffic is a high-octane adventure, where one gets the chance to continually sharpen one's skills, techniques, and wits, never quite knowing what might happen next. Bissell (2018), in

Figure 1.2 Traffic-themed graffiti in District 2, HCMC. Copyright 2019 by G. Wyatt.

the introduction to his recent book on commuting practices, points out that the word *commute* comes from the term *commutation*, whereby multiple single railway passenger tickets were *commuted* into a single payment railway pass. To *commute* is to be subsumed into Fordist efficiencies and mundane modes of moving the masses: seamless, uniform, and homogenized. The term 'commute' conjures images of a bus ride home from the office through unexceptional suburbs or being shunted through the concrete caverns of a subway system, isolated in a crowd of nameless faces with eyes transfixed to smartphone screens, like rabbits in headlights, frozen under fluorescent glare. In contrast to this, the unique structures of *mobilization* in HCMC were not born of the efficiencies of Fordist paradigms, but, instead, grew out of a complex, multi-layered mix of histories, beliefs, rural practices, technologies, and colonial influences.

The *style* of mobility in HCMC seems much less like *commuting* and much more like being caught up in a blizzard of constantly changing heterogeneous materiality, a dynamic, buzzing, zipping high-stakes chess game, and one Vietnamese scholar (Phạm, 2012) has made comment that the traffic in Vietnam is growing so fast that it is completely reconfiguring the traditional structure of society. My friend, on arriving at the bar, feels, I suspect, as though she has just been through some kind of war zone, not quite unscathed, in an environment where ethics are evolving parallel with emerging traffic practices. Most people who have lived some years in the city will say that the traffic has changed enormously in recent years and now seems more dangerous, perhaps because it is more heterogeneous, with many more cars, buses, and trucks, the speeds are faster, and the margins for error are becoming narrower. As cars increasingly dominate the physical space, often taking over the motorcycle lanes, an atmosphere brews in the traffic like a pressure cooker, an affective presence that has a profound influence on the emergence of new practices.

Sherburne, a scholar of the philosopher Alfred North Whitehead, quoting the French poet François Villon, asks, *Où sont les neiges d'antan?* "Where are the snows of yesteryear and how are they related to the present?" (cited in Sherburne, 1967, p. 253). Sherburne points out that Whitehead's answer to this would lead directly to God and that the past is preserved in the eternal realm, objectively immortal in his consequent nature. For Whitehead, it is God who sets up the initial conditions at the base of things, at the base of events, providing the seeds that inspire aims, cues, and relevance, resulting, over time – indeed, *creating* time – in outcomes of order that gel, if you like, with the unity of the universe. To view the HCMC traffic through Whitehead's ontology is to see it as constituted by events, varying in structural complexity, where God is seen as the divine orderer (Whitehead, 1978). Not to say that the kinds of order that result from these processes are pre-ordained or without autonomous creativity – creativity is central to Whitehead's scheme – but rather that the initial conditions set up by God limit or condition these creative processes

(Whitehead, 1978) and therefore mediate the kinds of forms that emerge. Whitehead's scheme relies on the preservation, in the eternal realm, of all that has ever been in actuality – such as in Vietnam's past – which then, through God, finds its way back into the primordial processes that create the actual universe. However, whilst the traffic in HCMC might be seen as constituted by events, it is not the events themselves that endure over time – when an event is gone, it is gone (Whitehead, 1920) – but it is enduring forms that go on to inspire novel future outcomes. For Whitehead, God is viewed as the only possible means by which the past may be given in the present, because he endures eternally and exists outside of time, and therefore, God *is* the ontological ground for the "somewhere" of eternality (Sherburne, 1967, p. 252).

Whitehead's process philosophy provides illuminating ways to view the complex processes and relationships through which eternal possibilities become concretely actualized. However, the existence of the entity 'God' in this scheme remains problematic for some scholars (Sherburne, 1967), existing almost as a kind of *deus ex machina*, a necessity existing beyond the temporal realm, and therefore able to create time itself. To consider Whitehead's metaphysical scheme *sans* God is to raise a number of productive questions, such as by what processes and dynamics do enduring 'things' reach in from the past and affect contemporary events, causally speaking, what is the nature of such enduring forms, how are they passed on, and where they reside whilst waiting to play their part?

Since Whitehead's time, the science of complexity – the notion of which is still open to debate (Corning, 1998; Érdi, 2008) – has evolved frameworks, definitions, and characteristics of systems that are dynamic and constituted through nonlinear relations, such as urban traffic systems. One of the most important characteristics of complex systems is a profound sensitivity to their histories, which fundamentally influences how they evolve over time. The endurances that resonate through the complex system that is the traffic in HCMC, for example, may have their seeds in historical events, but like Whitehead's God, they inspire outcomes in the contemporary world in ways that are impossible to predict or even to causally connect. Vietnam's 'snows' of yesteryear are characterized by the kinds of chaos that evolve in conditions of war, and this is a past that also includes years of extreme poverty, famine, and the complexities of colonialism. The question arises as to how this unique history might live on in the complex nonlinear dynamics that contribute to contemporary practices in Vietnam, for example, in its urban traffic. This is not to say that Vietnamese people *consciously* dwell in this history of war, but neither do they forget. There are many reminders of this history that are formally instituted in daily life, such as in commemorations of past events or in the names of streets, but HCMC is a young city, with much of its population born after the war (in fact, the population of Vietnam itself has the median age of 30 (CIA World Factbook, 2017), and, therefore, is future-oriented). Nevertheless, the forms that arose from this past linger on in subtle and complex ways, inspiring new structures, atmospheres and affects, relationships, and patterns of practices.

The HCMC traffic system, like an open blackbox, offers significant opportunities to observe how such an enormously complex and dynamic assemblage creates itself over time, responding to needs and evolving via a potpourri of often competing agendas, plans, and aims. At the level of the individual traffic user, this contribution occurs, not through conscious reflection and deliberation (who has the time for this?), but by way of entities engaged in what Dreyfus calls *absorbed coping* and what Noë (2012) calls *unthinking attunement* (p. 7).

HCMC traffic users are continually absorbed in kinetic forces, entangled in unseen enduring structures and reacting to ephemeral affective atmospheres, coping as best they can, often in extremely difficult conditions. As HCMC goes, in the words of Thrift (2008), "howling into the unknown" (p. 114), the nature of its traffic seems driven by nothing more than a "being-toward-movement" (Sloterdijk, 2006, p. 38), to exist is to move, as infrastructural forms conform to the emergent outcomes of the traffic itself, a kind of *kinetic materialism*.

Katz (1999), in his book *How Emotions Work*, says:

> Although people are intimately familiar with and vividly touched by the forces that shape their lived experiences, the critical shaping processes work like habitually worn eye glasses: they are transparent to the person even while, by structuring perception, they critically influence the person's response to his/her environment. (p. 8)

Figure 1.3 Sometimes things just need to be transported quickly and inexpensively. Copyright 2016 by G. Wyatt.

To engage with the traffic in HCMC can never be half-hearted, and it requires commitment, faith, belief, skill, trust, concentration, more than a little nerve, and, perhaps, the ability not to critically think about it too much. Like water for fish, we swim in fluid infrastructures of felt intensities and affective atmospheres, and no matter how intense the experience, as riding a motorbike in HCMC testifies, any practice might still become mundane, even if a *mundanity* of intensity.

In conjunction with Vietnam's unique historical legacies, as the country continues to open out to the world, other contemporary forces of modernity and mobilization also rush in, forces described by Giddens (2012) as "like being aboard a careering juggernaut rather than being in a carefully controlled and well-driven motorcar" (p. 53). Atkinson (2015) describes the atmosphere in HCMC as "a kind of euphoria where much of the population feel like they have arrived in the modern consumer world" (p. 857), but, as the philosopher Sloterdijk (2006) says, there exists military connotations, even in the term 'mobilization' itself. Sloterdijk (2006) states that "kinetics is the ethics of modernity" (p. 37), which raises the question as to whether the kinds of ethics and practices emerging in HCMC traffic via strategies of *absorbed coping* are, perhaps, emerging out of what Sloterdijk (2006) calls "moral kinetic automatism" (p. 38), reflective of the essence of *mobilization* itself, the modern form of which Sloterdijk describes as a "category of a world of wars" (p. 40). Truitt (2008) describes HCMC as a city "besieged" by motorcycles (p. 3), and news sites report vehicles fighting for space on the streets (Khanh An, 2015) and an increase in fights, sometimes using weapons, breaking out as a result of traffic incidents (Tran, 2014). Nghiêm Cường, a Vietnamese reporter who has broken his arm twice in two separate traffic accidents, states:

> If you can drive a car or bike in Vietnam, you can drive anywhere in the world. . . . We have rules but most people don't follow them and just ride the way they like. . . . Crashes here are very frequent. I see them every day. (Bryant, 2014, para. 1)

In this traffic system, ethics and morals do not exist in quiet contemplative ideological vacuums; we are not monks on motorcycles (except for the monks, who are, in all probability, actually monks on motorcycles). Instead, they emerge from out of – and, in fact, are part of – contingent concrete realities; yet, despite the obviousness of this statement, we continue to believe the state of affairs to be the other way around: that our attitudes and ideologies come first, and that it is independent, autonomous conscious thoughts that control and drive our behaviour.

Our individual destinies are profoundly influenced by, and connected to, the destinies of our urban environments, upon which it seems we have little control. As an individual being, I have my own hopes and goals, but this individual autonomy is limited (in the same sense that Whitehead's

God limits the fundamental creative processes that create nature) by the intrinsic programs that constitute me as a biological being, the systems I am embedded in, and the relationships that give meaning to my existence; I am not, in any real sense, a free and clear agent. Merleau-Ponty (1968) said, "it is not I who makes myself think any more than it is I who makes my heart beat" (p. 221). My autonomy as an individual, my actions, my feelings, and – as Merleau-Ponty points out – the *direction* of my thoughts are inseparable from the programs, modes, affordances, and 'aims' of systems, networks, protocols, and programs, the intimate relationalities that form the inhuman and nonhuman collectives within which I am a component. I drive, therefore I am, and my intentions and goals are intrinsically intertwined with the mobile infrastructure that we call 'driving a car' or 'riding a motorbike.'

Phenomenology has an uneasy and entangled relationship with metaphysics, but the way into the latter may well be by way of the former, just as much as the way into the former, be via the latter:

> Phenomenology, as an analysis of the origin of the world, is thus anything but a glorification of man as master of himself. On the contrary it shows him, by its results, as bound by, and under the authority of, a world order; and thus, metaphysics wins back that function which, from its nature, it has always had, and must have: the function of *reminding man of the bond which ties him to this order.* (Landgrebe, 1949, p. 205)

Figure 1.4 Child seats for motorcycles in parking area of apartment building. Copyright 2019 by N. Huynh.

References

Atkinson, A. (2015). Asian urbanisation. *City, 19*(6), 857–874.

Bissell, D. (2018). *Transit life: How commuting is transforming our cities*. Cambridge, MA: MIT Press.

Bryant, J. (2014, April 21). Risky road ahead for Vietnamese motorcyclists. *UQ in Vietnam*. Retrieved from http://uqinvietnam.com/2014/04/21/getting-a-handle-on-traffic-in-ho-chi-minh-city/

CIA World Factbook. (2017, October 27). *Vietnam*. Retrieved from https://www.cia.gov/library/publications/the-world-factbook/geos/vm.html

Corning, P. A. (1998). Complexity is just a word!. *Technological Forecasting and Social Change, 59*(2), 197–200.

Edensor, T. (2010). Introduction: Thinking about rhythm and space. In T. Edensor (Ed.), *Geographies of rhythm: Nature, place, mobilities and bodies* (pp. 1–18). Farnham, England: Ashgate Publishing, Ltd.

Érdi, P. (2008). *Complexity explained*. Berlin, Germany: Springer.

Giddens, A. (2012). *The consequences of modernity*. Cambridge, United Kingdom: Polity Press.

Hosinski, T. E. (1993). *Stubborn fact and creative advance: An introduction to the metaphysics of Alfred North Whitehead*. Lanham, MD: Rowman & Littlefield Publishers.

Katz, J. (1999). *How emotions work*. Chicago, IL: University of Chicago Press.

Khanh An. (2015, October 3). Ho Chi Minh City may become less livable due to unsolved traffic problems: Experts. *Thanh Nien News*. Retrieved from http://www.thanhniennews.com/society/ho-chi-minh-city-may-become-less-livable-due-to-unsolved-traffic-problems-experts-51973.html

Landgrebe, L. (1949). Phenomenology and metaphysics. *Philosophy and Phenomenological Research, 10*(2), 197–205.

Merleau-Ponty, M., & Lefort, C. (1968). *The visible and the invisible: Followed by working notes*. Evanston, IL: Northwestern University Press.

Noë, A. (2012). *Varieties of presence*. Cambridge, MA: Harvard University Press.

Phạm, N. T. (2012). *Đặc điểm giao thông Việt Nam từ góc nhìn văn hóa* [Traffic features in Vietnam from the cultural perspective]. Retrieved from http://vhnt.org.vn/tin-tuc/y-kien-trao-doi/27696/dac-diem-giao-thong-viet-nam-tu-goc-nhin-van-hoa

Sherburne, D. W. (1967). Whitehead without God. *The Christian Scholar, 50*(3), 251–272.

Sloterdijk, P. (2006). Mobilization of the planet from the spirit of self-intensification. *TDR/The Drama Review, 50*(4), 36–43.

Thrift, N. (2008). *Non-representational theory: Space, politics, affect*. Abingdon, England: Routledge.

Tran. H. (2014, June). Shocking fights occurring because of traffic accidents. *Vnxpress*. Retrieved from http://vnxpress.net

Truitt, A. (2008). On the back of a motorbike: Middle-class mobility in Ho Chi Minh City, Vietnam. *American Ethnologist, 35*(1), 3–19.

Whitehead, A. N. (1920). *The concept of nature: Tarner lectures delivered in Trinity College*, November 1919. Cambridge, United Kingdom: Cambridge University Press.

Whitehead, A. N. (1978). *Process and reality: An essay in cosmology (corrected edition)*. New York, NY: The Free Press. Originally published in 1929 by Macmillan.

2 Towards a metaphysics of experiential complexity

During the early evening of Tuesday, September 15, 2015, I found myself in, what is for Ho Chi Minh City (HCMC), the perfect traffic storm: heavy rain that lasted for hours, coupled with a flood tide, both occurring during the formidable chaos of peak-hour traffic. It had started raining heavily in the late afternoon, and it seemed a smart move to wait it out before attempting to ride my motorcycle home from work. When the rain seemed to abate, I donned my poncho-style Vietnamese raincoat and cautiously rode across town for home. Nearing the large Hang Xanh intersection on Dien Bien Phu Road in Binh Thanh district, a crowded and chaotic intersection at the best of times (an intersection, referred to with a certain fearful reverence by one long-term expatriate as 'the circle of death'), the traffic began to bunch together and, within minutes, I found myself stuck deep in a traffic jam. This is a wide road with many lanes and an overpass, all of which were crowded with wet steel, plastic, rubber, and raincoat-clad, anonymous figures. Most of the main thoroughfares leading towards this intersection had been transformed – at the time I was unaware of this – from roads, into surging rivers of floodwaters. This intersection, having been spared the flooding, therefore presented itself as some kind of Mecca, towards which buses, cars, trucks, bicycles, and thousands of motorcycles all made their way, like rats scurrying from a sinking ship.

HCMC sits on low-lying land and is surrounded by rivers, and water drainage has become increasingly problematic in certain areas. This situation is complicated by the fact that the flood-prone areas shift and move from one week to the next, as a result of changes to the environment wrought by the enormous number of construction projects throughout the city. Whilst the government is attempting to address the flood problem, such as through the building of small-scale flood barriers, it is a complex problem with many factors, and the increased flooding, according to one scholar, threatens "to render whole urban neighbourhoods unliveable" (Atkinson, 2015, p. 857).

Flooding occurs in HCMC from tidal currents and heavy rain and is compounded by insufficient drainage infrastructure. Tidal currents are connected with the phases of the moon, so, interestingly, the moon is an important actor in the traffic system, and one I often consult with a quick glance upwards when thinking about my immediate travel plans. Flooding creates all kinds of additional difficulties in what is an already challenging traffic environment, and one traffic user spoke of having to "swim to home," explaining how flooding is the most terrible aspect of the traffic experience in HCMC. Floodwater on the roads decreases the amount of usable space and it forces motorcycles into car lanes, cars into motorcycle lanes, and makes it difficult for buses to pick up passengers. These floodwaters can also be extremely toxic, and the waves caused by larger vehicles can destabilize motorcycles, causing the rider to fall into the waters, an outcome that is potentially extremely dangerous for one's health.

Figure 2.1 Coming home from work through flooded streets, District 2, HCMC. Copyright 2016 by G. Wyatt.

Figure 2.2 Motorcycles riding through flooded streets. Copyright 2016 by G. Wyatt.

Figure 2.3 Flooded streets, District 1, HCMC. Screenshot from video footage taken from government bus, 2016.

Photographs of the traffic jam at Hang Xanh intersection appeared on-line the following day, but these images, mostly taken from above the scene, whilst impressive in terms of displaying the sheer scale and spectacle of the event, failed to communicate the atmosphere and experience of being stuck deep within it. I remember the sky was grey, and there was still a drizzling mist, heavily tainted by accumulated exhaust fumes. For a while, I sat on my motorcycle just a few metres from the rear of a large black–green govern-ment bus, but I was unable to manoeuvre away from it. Black smoke issued from its exhaust pipe, so thick, that it was really more liquid than gas, and with the exhaust fumes being emitted from the other hundreds of vehicles in the tight space, the air became so polluted that I feared I might pass out. I sat somewhat soggy on my motorcycle, worried about what was to come, unable to go forwards, backwards, sideways, up or down, trapped in the affordances of an environmental–technical–cultural milieu. *What if there was a fire*? I thought. What if we all actually *had* to move? It was clear that we were at the mercy of something bigger and more powerful, entities on par with each other, all reduced to the same level, locked into the same immo-bile structural complexity, all mere components, organic and mechanical together: the many had become one and the one had become many.

I looked around the scene, trying to catch the eye of other riders in order to ascertain the general mood, and I was struck by the fact that no one directly engaged with anyone else. We were intimately connected and yet, strangely, we remained coldly isolated from each other (intimate isolation). Luckily not mass hysteria – a trait, fortunately, uncharacteristic for the Vietnamese – but certainly mass *denial*, as though the events surrounding us were somehow not really occurring, and this created an atmosphere that was as surreal as it was dislocating. Like being in the eye of a hurricane, it was peaceful, yet filled with tension and expectation, as intense kinetic energies were captured like an overly energetic dog that had wrapped his lead too many times around a pole. Even the slightest movement was impossible and if a space of even a few centimetres opened up, a vehicle pounced on it, only serving to crush the mass together tighter.

Some months later, I read a paper entitled "The Tragedy of the Commons" by Hardin (1968), in which he discusses the idea that some situations are beyond the scope of scientific or technical solutions, and I was immediately reminded of my time spent in this traffic jam. Hardin draws on a fable in which each herdsmen, sharing a common pasture, adds one more animal to his herd, and then another and another, until the pasture is overgrazed, resulting in disaster for the whole society. Hardin (1968) writes, "ruin is the destination toward which all men [sic] rush, each pursuing his own best interest in a society that believes in the freedom of the commons" (p. 1244). This traffic jam was our pasture, our commons, a complex emergent outcome beyond mere technical solutions.

A term often heard in reference to traffic behaviour in HCMC is the Vietnamese term *chen lấn*, a negative term that loosely means to 'jostle,' as in when a person or a thing tries to expand their zone by moving forward into someone or something else's zone. *Chen lấn* is a kind of 'me first' attitude that also manifests in a style of movement, the kind of jostling behaviour that would be more commonly observed in a crowded market place full of pedestrian shoppers, all anxious to get a bargain, for fear of missing out. It is often associated with selfishness, self-centredness, losing patience, and disregard or lack of awareness for others. As one HCMC traffic user observed, "whenever they see a space on the road, they would take it to proceed. They overtake to go first . . . Vehicles do not follow the traffic rules and everyone wants to go first and overtake carelessly."

Chen lấn might be viewed as an enduring form, at once both attitude and action, a form that would continue to permeate all manner of technical solutions (see Figure 2.4).

In the rare stillness of this traffic event, that which was usually backgrounded amid the manic instrumentalism of simply getting to where I needed to go, became foregrounded, and a strange new tableau formed before me, not of materiality as substance, but of *relations as substance*, the *substantial* existence of relationalities imbued with the form of *chen lấn* mixed with *technicity*; in other words, the *eventness* of the event. This was neither 'traffic' nor 'commuting,' but something else, a new kind of nature perhaps, our new forest, an *Anthropocenic* implosion in slow motion, either way, it was clearly something that needed a new set of abstractions beyond algorithms and a new framework or approach towards better understanding it.

With this experience in mind, this book aims to explore the aesthetic, the style, and the unique character of the HCMC traffic, not as a timeless and

Figure 2.4 Cars quickly move into available spaces, resulting in immobility for all.
Copyright 2017 by G. Wyatt.

static essence, but how character is achieved and produced as an ongoing and dynamic project; in Candea's (2008) words, "how essence is put together" (p. 210). Mata (2016), in the paper entitled, "A Phenomenological Investigation of the Presencing of Space," writes, "phenomenology begins in silence" (p. 26). He describes how *phenomenological seeing* emerges through quiet contemplation, using examples such as his own childhood experiences and leisurely strolls around the Golden Pavilion in Kyoto. To some extent, this is true, as was the case with my traffic jam experience, but this is not the usual *givenness* of the HCMC traffic, and would a stroll around the Golden Pavilion, amongst crowds of tourists, actually be an experience of quiet contemplation?

The beauty of phenomenology, as Moran (2000) has noted, is that it always resisted becoming a system or even a method embedded in any rigid set of dogmas. Instead, phenomenology, as is the case here, may be viewed as a kind of philosophical orientation, more than a method, a way into thinking about essences, perception, and consciousness (Merleau-Ponty, 2005), describing, rather than explaining (Ehrich, 1999), and attempting to get a sense of the varying structural relationalities of experience. Moran describes phenomenology as:

> A radical, anti-traditional style of philosophising, which emphasises the attempt to get to the truth of matters, to describe phenomena, in the

broadest sense as whatever appears in the manner in which it appears. . . . As such, phenomenology's first step is to seek to avoid all miscon- structions and impositions placed on experience in advance. (Moran, 2000, p. 4)

In this spirit, the ethos of this book is a phenomenology, not merely of quiet contemplative and reflective thought, but one that begins in noise and movement, more in the manner in which the phenomenon itself appears – its *givenness*, to use Husserl's term – described from the inside, a view from somewhere, rather than from nowhere, deep within concrete matters, rather than from a drone flying above, or through the eyes of God.

This book intends to add to the various and multidisciplinary research areas concerned with phenomenology, affect, affective urbanism, and also those represented under the broad banner of "mobilities," a wide ranging body of work that concerns itself with notions around movement, flow, and the liquidity of people, objects, information, communication, and mobility capital, a discipline that encompasses, amongst others, areas such as trans- port geography, cultural geography, human geography, and the anthropol- ogy of circulation, migration and tourism (Sheller, 2014). In conjunction with the "mobilities turn," occurring from about the year 2000 onwards (Grieco & Urry, 2011; Sheller, 2014), this research space saw increased attention on ur- ban traffic, especially studies of a sociological nature, of particular note, from scholars such as Beckmann (2004), Dant (2004), Featherstone (2004), Miller (2001), Sheller (2000, 2005, 2006), Thrift (2004), and, of course, Urry (2000, 2004, 2006, 2010) who wrote of "assemblages," "scapes," "flows," and "networks," also incorporating notions of complexity (Urry, 2005).

Through these studies, urban traffic began to be viewed more in terms of its sensory and emotional dimensions, for example, and of particular note, Sheller's study, "Automotive Emotions: Feeling the Car" (2005), which explored the culture of automobility as "affective and embodied relations between people, machines and spaces of mobility and dwelling, in which emotions and the senses play a key part" (p. 221). This more aesthetically oriented approach is, in Sheller's case (2005), an attempt to move research beyond the mere "statistical quantification of individual preferences, atti- tudes and actions" (p. 222). In this way, Sheller takes into account the emer- gent effects and affects of the *combined* affective, cognitive, and physical dimensions, through the formulation of a theoretical framework she calls an "emotional sociology of automobility" (p. 223). However, Sheller's study, like many other sociological studies of urban traffic, tended towards rep- resentations of traffic in a more universal form, rather than a specific place- based phenomenon. In fact, the dominant trend of the studies that grew out of this research movement was the formation of a universal form of traffic that retained a Euro-American bias, equating urban traffic and automo- bility with *car* traffic, rather than a traffic mix made up predominantly of motorcycles, as is the case with the traffic in HCMC, and analyses of the kinds of dynamics present in a traffic system that evolved with motorcycles, such as in Vietnam, are all-but non-existent in this literature. Also, many of these studies, whilst viewing traffic in terms of its aesthetics, sought to understand human–object relations in traffic through more semiotic, symbolic, and *representationalist* ways, often focusing on values, attitudes, and preferences within sociologically oriented conceptual research themes such as identity and consumption (for example, Carrabine & Longhurst, 2002; Dant & Martin, 2001; Edensor, 2004). In a similar way, Taylor (2003), in his study entitled "The Aesthetic Experience of Traffic in the Modern City," wrote of the aesthetic evaluation of urban places, based in sensorial experience, but beginning with value judgements residing in conscious thought, without attendance to a phenomenological account of what might be *antecedent* to such evaluations of atmospheric spaces.

On the other side, growing out of the science of complexity and transportation research fields, other less sociologically oriented approaches to traffic studies attempt to engage with the complexity of traffic and are often embedded in paradigms of mathematical physics, algorithms, and the power of computation. These studies are often much more technologically oriented and utilize models of representation, using the tools of computer modelling and simulations, that separate relations from entities or construct relations different in nature from the experiential subjective traffic worlds, thereby manifesting abstract traffic realities that may be considered, according to one scholar Hayles (1999), as worlds of their own.

This book aims to supplement these studies and to explore how we *feel* driving, rather than how we feel *about* driving. As Galloway (2013) has said, "for phenomenology, the solution to any problem is always found in the irreducible authenticity of the feeling subject, never the dry calculations of math and science" (pp. 361–362). Rather than begin with a universal representation of traffic or of traffic as a mathematical construction, this book seeks what is present (even if present through absence), and what is *given*, by way of the relations between experience and perception, and style and character in a concrete and *particular* traffic system, at the points where ontology, experience, perception, and materialism meet. This is, therefore, a phenomenological account of specific experiences, specific articulations, and fluid and dynamic assemblages, rather than static universal formulations of the 'car' or of 'traffic.' As an exploration into the forces, dynamics, and processes that influence, form, and constitute the traffic system in HCMC, the focus remains on relations in a way that keeps the event intact, so that, as Harman writes, "every relation must be regarded as a substance in its own right" (Harman, 2005, p. 94). The traffic is then viewed in terms of series of fluid processual relational events, from which emerge particular enduring forms or modalities, or in Whiteheadean terminology, *prehensions*. From simple and complex events, forms and objectified abstract endurances coalesce and coagulate into nexus, societies, communities, amalgamations, networks, and 'things,' held together by relationalities that adhere to particular logics, character, and style. Somewhere in between the event-present and the emergence of enduring objects, patterns, and forms, lay our *experiential complexities*, or what Latour (2008b) calls "matters of concern," forming the porous borders between the *pure mobility* (Stengers, 2011, p. 74) of pre-sensorial, pre-conscious, and pre-intellectual awarenesses and the conscious cognition of significance, meaning, and value judgement.

As *feelers* and *feelings*, subjects and objects emerge from these kinds of affective modes of being, co-constituting our traffic worlds and our sense of self. In this way, this book seeks to investigate how, through processes of awareness and perception, abstractions such as objects, time, and space, as well as relevance, meaning, and significance, emerge out of fluid, dynamic, open-ended potentiality. In other words, this is not a book about traffic, *per se*, but about the nature of our experience of nature, and, perhaps, something of the nature of nature itself, as told through the traffic in HCMC as its object. It is an attempt to describe the forms and relationalities that constitute the affective spaces, atmospheres, and character-laden relationalities within which traffic users dwell, which they entangle with, and from which emerge 'subjects' and more permanent abstractions. These fluid media that flow between these states are the emergent trans-intensities (Anderson, 2009) of affect and atmosphere, which are the "becomings" (Deleuze & Guattari, as cited in Anderson, 2009, p. 78) "experienced in a lived duration that involves the difference between two states" (Deleuze, as cited in Anderson, 2009, p. 78). The open nature of experiential relational complexities, from which the closure of abstractions emerges, is non-representational, described by McCormack (2010) as "a kind of distributed, immanent field of sensible processuality" (p. 202) that, importantly, is "without transcendent reflection" (p. 202).

Given that the experience of traffic – a quintessentially non-representational, aesthetic, and sensory experience – is, for most, the central feature of the experience of modern cities, and that 'place,' to a very large degree, is characterized by the nature of the mobilities that course through it, the relative dearth of writing on the aesthetic and affective dimensions of traffic is surprising (Taylor, 2003).

Therefore, this book takes up the following call from Thrift, who suggests:

> Yet it is still possible to write of a rich phenomenology of automobility, one often filled to bursting with embodied cues and gestures which work over many communicative registers and which cannot be reduced simply to cultural codes . . . and [to] understand driving as both profoundly embodied and sensuous experiences. (2008, p. 80)

In doing so, this book attempts to develop a *metaphysics of experiential complexity*, which, despite the rather grandiose title, might provide a framework, an orientation, and a practical approach in the analysis of the experiential and meaningful events that constitute subjects and objects as intimate amalgamations in complex urban environments. In summary, this approach prioritizes the relational particularities that constitute the series of "thick" (Nail, 2019, p. 372) *event-presents* that fold and merge the past, present, and future in a fluid medium of rhythmic, atmospheric, and affective time-spaces. It seeks to reveal, through the traffic of HCMC, the destructive creative nature of events, and the objectified abstract endurances that emerge from them, such as materiality and physicality, time, space, emotions, objects, subjects, infrastructural complexes, and assemblages of patterns of practices.

Embedded in the notion of the *event*, this book also intends to develop the idea of atmospheres in traffic as substantial, yet fluid and ephemeral infrastructures that inspire aims and guide our actions. For Whitehead, the birthplace of nature is always the event, and, at its most fundamental, the *actual entity*, which Whitehead describes as the most fundamental building block of the universe, an entity that is both process and thing. Such an analysis requires twin perspectives, two eyes and ears in very different places, on the macro and the micro, yet without discrimination between the two, to see the micro *as* the macro. Events are always particular, unique, and contingently present, but they lead us out into broader generalities as well as into the future and the past. What is the phenomenality of driving, if not a series of events, from which temporality, spatiality, and materiality construct an entire world, changing "drastically, the whole style of an existence" (Dastur, 2000, p. 182). The act of driving a car, as a mode of being and a series of events, is not something that occurs *in* a world, but, rather, is a happening that *creates* world, the constitution of never-quite-resolved openness, essences, contingencies, and accidents (Dastur, 2000).

An affectively experienced event, then, is like a rock pool, containing an ocean of presence and absence, materiality and ideality, definiteness and indefiniteness, clarity and vagueness, and singularity and generality (Anderson, 2009). Its particular and contingent relations act on the past and feed the future, reflecting an ocean filled with the oceans before that. As Gould (1990) wrote, in his book on the adventures and enormous implications of finding fossils:

> God dwells among the details, not in the realm of pure generality. We must tackle and grasp the larger, encompassing themes of our universe, but we make our best approach through small curiosities that river our attention-all those pretty pebbles on the shoreline of knowledge. For the ocean of truth washes over those pebbles with every wave, and they rattle and clink with the most wondrous din. (p. 51)

The HCMC traffic system is, at once, artwork and assemblage, a rich fabric of multiplicities, constituted of different issues, concerns, and experiences, with consistently appearing logics and cohesive forms and stratifications flowing in and out of its porous borders into other domains of life. Traffic vehicles regularly overflow beyond actual roads, as motorcycles – and sometimes, even cars – travel along sidewalks in search of a less obstructed path, and motorcycles are often parked in the kitchen or living room, perhaps right next to the television set. The sounds of construction trucks – forced to drive at night due to traffic restrictions in the city – enter one's bedroom like the sounds of wild animals beyond the cave walls. Motorcycle helmets are often worn as hats, in a more general sense, even when there is no motorcycle in view. Hot noodle soup, steaming in bowls, along with chopsticks, is precariously delivered on motorcycles, balanced on trays, and carried waiter-style, leaving the rider/waiter only one hand to operate the bike in an acrobatic feat of balance and agility. The motorcycle is embedded into almost all levels of practices in Vietnam, and the infrastructure has grown up around them, so that riders are able to do market shopping and a multitude of other tasks, including fishing, dating, sightseeing, sleeping, window shopping, eating, and drinking without even having to leave the motorcycle seat.

For Whitehead, everything is what it is, only through its connections to everything else. Such an ecological ontology necessitates a methodology that attends to the multiple modes of connections and relations – moving

Figure 2.5 Motorbikes are often seen carrying all kinds of loads, but rarely another motorbike. Copyright 2019 by G. Wyatt.

from standpoint-to-standpoint – between humans and nonhumans from which meaning, significance, and relevance emerge for all actors in traffic. Given the interwoven nature of the phenomenon, the non-reductionist approach presented here seeks to keep the relations of the particular performance intact and so focuses on the event. When the event is given ontological centre stage, perspectives shift in ways that allow paradoxes, such as between immobility and mobility, revealing modes that neither subject nor object can solely lay claim to because the event, which, itself, provides a particular standpoint or perspective, also belongs to "the great impersonal web of events" (Stengers, 2011, p. 65) that bridges the eternal realm of the possible, with the world of the actual. In this way, every single actual entity, either through incorporation or rejection, presents everything that has ever been. A green traffic light is what it is, due to its not being red (or blue or purple, or, for that matter, a cat), and so such an event stretches out into the endless realm of comparison and of what could have been but was not actualized as such.

Tarde (2000) warns of

> The error of believing that, in order to see a gradual dawn of regularity, order, and logic in social phenomena, we must go outside of the details, which are essentially irregular, and rise high enough to obtain a panoramic view of the general effect. (p. 75)

It is for this reason that this approach seeks the particular over the universal, whilst also viewing the particular *as* the universal, formulating essences and abstractions that keep their meaningful relationality and rich entanglements intact, amplifying the contingent modalities of being.

This more aesthetic, affective, and processual approach to the evolutionary processes at work in HCMC contrasts with many dominant traffic research approaches, which are so often technological-solutionist in orientation, technologically motivated, and embedded in mathematical algorithms and paradigms. Smart cities, intelligent transportation systems, and technology, in general, are often viewed as a fix-all solution to many urban problems. However, the dominance of such research paradigms, especially those that tend towards a mathematicization of nature, often serve to create their own realities and ontologies, relationally different from concrete experience, and fall prey to what Whitehead calls the *bifurcation of nature*.

In developing this notion of a metaphysics of experiential complexity, I have drawn upon various theoretical frameworks, ideas, and scholars in order to create an approach that I consider most suited to the character of the particular phenomenon under analysis, namely the HCMC traffic experiences, including:

- The existential spatiality and phenomenology of Heidegger,
- The embodied phenomenology of Merleau-Ponty,

- The process philosophy of Alfred North Whitehead,
- Affect theorists such as Brian Massumi and Erin Manning,
- Bruno Latour and Actor-Network Theory (ANT),
- Object-Oriented Ontology, such as from Graham Harman,
- Nigel Thrift's Non-Representational Theory, and
- The radical empiricism of William James.

Whilst some might find ontological incompatibilities that prohibit the inter-weaving of all of these various perspectives, the goal here is not to pit one against the other, but to complement and supplement, that each may add new dimensions or enlighten one another. As one scholar wrote:

> It is necessary less to dogmatically oppose a phenomenological or Hei-deggerian orthodoxy to Whitehead's metaphysical enterprise, than to seek to describe what gives itself to thinking, the experience in its fun-damentally sensible dimension, the appearing in its sensible inscription, being as sensitive being. (Robert, 2005, p. 367)

Some commentators have suggested that Whitehead's process philosophy lacks moral and political dimensions, especially in terms of the social impli-cations of the 'decisions' made by actual entities (Sherburne, 1983) or, what Grange calls "the consequences of finitude" (as cited in Rice, 1989, p. 184). Therefore, Whitehead's metaphysics benefits from cross-fertilization with, for example, the more existential character of Heidegger's phenomenology. Whitehead's account of human experience is sometimes seen as just another instance of the general scheme of metaphysical principles (Shade, 1995), while existentialism takes into account the significance and uniqueness of the questioner for whom the issue of *being* is an issue, a significant ontologi-cal departure point in itself (Rice, 1989). Indeed, very similar criticisms have been made of ANT, especially in its inability to account for the *social* impli-cations of networked agency through technical decisions (Walsham, 1997). Sherburne has suggested that an integration of Whitehead's process philos-ophy and the phenomenological existentialism of Heidegger would create a synthesis, equivalent of a Copernican Revolution, in agreement with Gier (1976), who notes the "substantial parallels" (p. 211) between Whitehead and existential phenomenology. Yet, despite these promising fusions, studies that further a dialogue between Heidegger and Whitehead are rare (Shade, 1995), and one scholar even suggests that "No dialogue has truly taken place between phenomenology and Whitehead" (Robert, 2005, p. 366).

When viewed from the more existential side, Heidegger's ontology, which is predominantly characterized by a transcendentalist perspective, is often criticized for its subjective bias, placing, as it does, Dasein at the centre of all enquiries, a limitation that would benefit from the more symmetrical ontol-ogies of Whitehead and ANT. Whitehead begins his investigations with hu-man awareness simply because this is the aspect of nature most accessible to us, but, unlike Heidegger, Whitehead *decenters* the human and approaches the human experience as an opportunity to better understand the nature of nature and the structures of experience, broadly expanded to include non-human realities. In a sense, Whitehead takes the subjective *givenness* of the phenomenal experience and extends it out into the world to include nonhumans and inanimate entities, redefining the notion of the Cartesian subject as well as what it means to 'experience' some *thing*. Whitehead's 'subjects' need not necessarily be of a human or idealist foundation because the fundamental processes of an experiencing subject are the same, even if the entities doing the 'experiencing' are non-sentient beings, such as a stone or a motorcycle tyre. However, though the processes are the same, it is only complex entities that achieve mental activity through the limitations of abstraction, as "consciousness is only the last and greatest of such elements by which the selective character of the individual obscures the external totality from which it originates and which it embodies" (Whitehead, 1978, p. 15).

Dastur (2000) makes comment on how Merleau-Ponty could claim neither a realist, nor an idealist solution to the problem of accounting for time and, for this reason, she says, "we should not oppose phenomenology and the thinking of the event. We should connect them; openness to phenomena must be identified with openness to unpredictability" (p. 178). This broadens the scope and possibilities of phenomenology as a tool and as an orienting perspective on complex phenomena, remedying some of the criticisms of phenomenology, such as from Latour (1999), who claims that it "deals only with the world-for-a-human-consciousness" (p. 9). Whilst ANT also promotes perspectives that move us from standpoint-to-standpoint, it is the focus on the event that provides the kinds of openness that encompasses relations between the eternal of all possibilities and the closure of the concrete actual.

It is certainly no secret that Latour, post ANT, has "deep allegiances" (Weber, 2016, p. 522) with Whitehead, borrowing and further developing some of Whitehead's fundamental ideas, particularly the problem of the bifurcation of nature. This problematization, termed the 'bifurcation' underpins much of Latour's notions of 'matters of concern' (Latour, 2004, 2005, 2008a, 2008b), the 'Modern Constitution,' as described in his book *We Have Never Been Modern* (1993), and his more recent work on the modes of existence (2013). In fact, Latour (2004) has suggested that Whitehead's term, 'actual occasions,' describes exactly what he, Latour, means by his notion of 'matters of concern' (p. 245). All this suggests great possibilities for the development of a framework of enquiry that weaves together existential phenomenology, process philosophy, and network theories such as ANT.

Urban traffic systems are considered examples of complex adaptive systems, so it seems appropriate to incorporate some aspects borrowed from complex systems theories in order to understand the kinds of nonlinear dynamic interactions and relationalities that allow us to conceptualize certain kinds of systems and problems. Robinson (2005) has noted the ways in which Whitehead's ontology, together with Deleuze, "resonates with the emerging sciences of complexity" (p. 175), and he has drawn upon both Whitehead and complex systems theory towards the creation of what he calls "a metaphysics of complexity" (p. 159), a phrase that inspired the name of the approach presented here. These various ontologies all generally share an aversion to representationalism, conscious subjectivism, or stubborn idealism, instead, focussing on relations, connectivity, and interdependence, and, in many cases, on the vague pre-intellectual kinds of awarenesses that emanate from the past, that Whitehead (1978, 1985) calls *causal efficacy.*

Edensor and Jayne (2012) point out that urban theory, unlike other social sciences, has been slow to problematize and challenge dualisms such as 'West,' 'non-West,' 'developing,' and 'developed,' because urban theory has being dominated by research focused on "a small number of cities mostly located in North America and Europe" (2012, p. 2). They note that disciplines such as anthropology, biology, history, and geography were complicit in the proliferation of dualisms demarcating the notion of the 'other' in comparison to European identity. Within the genre of mobilities studies, whose scholarship on traffic has been almost entirely focused on Western versions of traffic, namely, car traffic, it would be easy for the present book to fall into a comparative us/them, Western/Asian, and motorcycle-traffic/car-traffic discussion of dichotomies (cars and motorcycles in HCMC exist in more complexly entangled ways than in dualistic ways). Instead, this book aims to attend to the givenness of the phenomenon, which requires, in line with Husserl's notion of the *epoche*, to "avoid all misconstructions and impositions placed on experience in advance" (Moran, 2000, p. 4) and to put aside all "scientific, philosophical, cultural and everyday assumptions" (Moran, 2000, p. 11) in order to return to the things as given in their contingent modes. Therefore, it makes little sense to construct theoretical realities around such dichotomies from the outset, only to spend the rest of the book attempting to dismantle them. Though contemporary Vietnam is a product of the many complexities arising out of its colonialist past, to view this phenomenon of traffic in HCMC through a theoretical framework borrowed from post-colonial studies is not the most appropriate avenue of enquiry, given the book's aims. The historical modes of enquiry embedded in many post-colonial studies are steeped in the language of dualistic identities, notions of the 'other,' the "borderline experience" (Bhabha, 2004, p. 296), and margins and demarcations of cultural difference. Perhaps it is due to, what some scholars have called, the "persistence of situated theories" (Edensor & Jayne, 2012, p. 2) that post-colonial studies expend energies problematizing these historical narratives of dualism within familiar themes, aims, and agendas. Even in problematizing these themes, as Chibber points out in his book, *Postcolonial Theory and the Specter of Capital*, post-colonial theory often, paradoxically, ends up endorsing the very things it aims to dismantle, namely Orientalism and Eurocentrism, due to an insistence on the universal as fixed and static (Birch, 2013).

Methodology

In addition to the present author's many years of experience in HCMC, this book also draws on a research project conducted in HCMC between 2015 and 2017 that focused on the experiences of a multiplistic sample of traffic users. These included Vietnamese and non-Vietnamese drivers and riders of cars, motorcycles, trucks, taxis (both cars and motorcyclists), government-operated buses, vans, and included traffic-centric occupations such as pedestrian lottery ticket sellers and delivery couriers. To capture data, these traffic users wore mounted GoPro cameras in order to gather audio-visual footage of their usual journeys in the traffic, such as from home to work (see Figure 2.6).

These users were then invited, along with another participant, for a video-viewing session. The pairs of traffic users present at the viewing sessions were selected in order to maximize differences in the experience of traffic, such as, for example, in size or materiality of their vehicles (truck driver paired with a motorcyclist), differences in cultural background, or paired based on their similarities (such as both being professional drivers). These users had the opportunity to view and discuss each other's traffic videos,

Figure 2.6 Composite of screenshots taken from video footage from different HCMC traffic users, 2016.

adding narratives, and both were able to control the video (stop, start, or pause). These sessions were videoed as both a screencast of what the participants were viewing and from the computer's inbuilt camera in order to show the reaction of participants as they viewed the footage. It should be noted, however, since this data was collected two years before the writing of this book, the introduction of so many new cars into the traffic system has further evolved the state of affairs in various different ways; this is a traffic system changing so fast, that even data only a few years old might be, if not obsolete, perhaps inaccurate.

The use of video as a research tool in this study is similar to studies by Jordan and Henderson (1995), who have used videotaping in conjunction with ethnographic work, a method they call "interaction analysis," which they describe as an "interdisciplinary method for the empirical investigation of the interaction of human beings with each other and with objects in their environment" (p. 39). As Jordan and Henderson (1995) note, the advantage of video is that it reveals the hidden rhythms and the often invisible, fleeting phenomena that serve to organize our lives. This method provides a way into social practices that are often automatic, hidden, and responsive to subtle shifts in the environment, and it is particularly useful for the analysis of routine practices and human interactions with artefacts and technologies, including verbal and non-verbal interaction. Raingruber (2003), who was guided by similar philosophical tenets to the present study, uses video in a similar way, a method she terms "video-cued narrative" (p. 1155). She reports that it helps participants (who, in this case, were outpatient nurse-therapists and their clients) to "re-collect, reexperience, and interpret their life world" (p. 1156). Raingruber (2003) sees this method as a valuable technique for phenomenology, which she says is fundamentally a method without techniques and, following Heidegger and Merleau-Ponty, is especially useful for accessing "relational, practice-based, and lived understandings" (p. 1155).

It is important to realize, however, that video is not a factual record of events any more than the participant's narratives are. It is a tool that allows us to focus on aspects of events that may otherwise go unnoticed as well as provide another standpoint or perspective for those events. The life of the video footage stretches out beyond the frame and is open-ended and ambiguous (Buur, Binder, & Brandt, 2000) to be used more as a stimulus or prompt for discussion, rather than an adjudicator of truth or fact.

The raw video footage from these user's traffic journeys and the data obtained from the viewing sessions (this data includes the screencasts of what the participants were viewing on the screen, the video showing their reactions to what they viewed, and the translated transcriptions of the viewing sessions, as the sessions were mostly conducted in Vietnamese language with a translator and research assistant present) were then compiled and synched together for later analysis by the researcher. The narratives of the participants, together with the GoPro video footage (the angle and perspective of which put the viewer deep in the traffic action), provided ways into the affective, sensory, and emotional dimensions of the traffic and often revealed the incorporation of embodied awarenesses and tacit knowledge and skills that the users themselves were often unaware of. Narratives still retain strong links to other kinds of awarenesses and continue to be productive ways into 'nature,' because they reveal traces of other submerged awarenesses (Pink, 2012). The video, together with the narratives, holds the event together – as much as possible, given the chaotic nature of this traffic system – in a non-reductionist way and the 'dissection' that occurred later in the viewing sessions can also be non-reductionist, more, as Massumi (2008) says, like "a revisory verbal echo of the perceptual déjà-vu of the semblance" (p. 25), as a holistic glossing over that makes sense of the semblance and can also place the event within a larger context.

References

Anderson, B. (2009). Affective atmospheres. *Emotion, Space and Society*, *2*(2), 77–81.

Atkinson, A. (2015). Asian urbanisation. *City*, *19*(6), 857–874.

Beckmann, J. (2004). Mobility and safety. *Theory, Culture & Society*, *21*(4–5), 81–100.

Bhabha, H. K. (2004). *The location of culture*. Abingdon, United Kingdom: Routledge.

Birch, J. (2013, April 21). How does the subaltern speak? *Jacobin*. Retrieved from https://www.jacobinmag.com/2013/04/how-does-the-subaltern-speak/

Buur, J., Binder, T., & Brandt, E. 2000. Taking video beyond 'hard data' in user centered design. *Proceedings of Participatory Design Conference 2000* (pp. 21–29). New York: CPSR.

Candea, M. (2008). Fire and identity as matters of concern in Corsica. *Anthropological Theory*, *8*(2), 201–216.

Carrabine, E., & Longhurst, B. (2002). Consuming the car: Anticipation, use and meaning in contemporary youth culture. *The Sociological Review*, *50*(2), 181–196.

Dant, T. (2004). The driver-car. *Theory, Culture & Society*, *21*(4–5), 61–79.

Dant, T., & Martin, P. (2001). By car: Carrying modern society. In J. Gronow & A. Warde (Eds.). *Ordinary consumption* (pp. 143–157). London, United Kingdom: Routledge.

Dastur, F. (2000). Phenomenology of the event: Waiting and surprise. *Hypatia*, *15*(4), 178–189.

Edensor, T. (2004). Automobility and national identity: Representation, geography and driving practice. *Theory, Culture & Society*, *21*(4–5), 101–120.

Edensor, T., & Jayne, M. (Eds.). (2012). *Urban theory beyond the West: A world of cities*. Abingdon, United Kingdom: Routledge.

Ehrich, L. C. (1999). Untangling the threads and coils of the web of phenomenology. *Education Research and Perspectives*, *26*(2), 19.

Featherstone, M. (2004). Automobilities: An introduction. *Theory, Culture & Society*, *21*(4/5), 1–24.

Galloway, A. (2013). The poverty of philosophy: Realism and post-fordism. *Critical Inquiry*, *39*(2), 347–366.

Gier, N. F. (1976). Intentionality and prehension. *Process Studies*, *6*(3), 197–213.

Gould, S. J. (1990). *Wonderful life: The burgess shale and the nature of history.* New York, NY: WW Norton & Company.

Grieco, M., & Urry, J. (Eds.). (2011). *Mobilities: New perspectives on transport and society.* Farnham, United Kingdom: Ashgate Publishing, Ltd.

Hardin, G. (1968). The tragedy of the commons. *Science, 162*(3859), 1243–1248.

Harman, G. (2005). *Guerrilla metaphysics: Phenomenology and the carpentry of things.* Chicago, IL: Open Court.

Hayles, K. (1999). *How we became post human: Virtual bodies in cybernetics, literature, and informatics.* Chicago, IL: University of Chicago Press.

Jordan, B., & Henderson, A. (1995). Interaction analysis: Foundations and practice. *The Journal of the Learning Sciences, 4*(1), 39–103.

Latour, B. (1993). *We have never been modern.* Cambridge, MA: Harvard University Press.

Latour, B. (1999). *Pandora's hope: Essays on the reality of science studies.* Cambridge, MA: Harvard University Press.

Latour, B. (2004). Why has critique run out of steam? From matters of fact to matters of concern. *Critical Inquiry, 30*(2), 225–248.

Latour, B. (2005). From realpolitik to dingpolitik: Making things public: Atmospheres of Democracy. In P. Weibel & B. Latour, (Eds.), *Making things public* (pp. 14–44). Cambridge, MA: MIT Press.

Latour, B. (2008). A cautious prometheus? A few steps toward a philosophy of design (with special attention to Peter Sloterdijk). In F. Hackne, J. Glynne & V. Minto (Eds.), *Proceedings of the 2008 Annual International Conference of the Design History Society* (pp. 2–10), Universal Publishers.

Latour, B. (2008b). What is the style of matters of concern. In *Two lectures in empirical philosophy.* Department of Philosophy of the University of Amsterdam, Amsterdam: Van Gorcum. Retrieved from http://www.bruno-latour.fr/sites/default/files/97-SPINOZA-GB.pdf

Latour, B. (2013). *An inquiry into modes of existence.* Cambridge, MA: Harvard University Press.

Massumi, B. (2008, May). The thinking-feeling of what happens. *Inflexions,* 1–40.

Mata, F. (2016). A phenomenological investigation of the presencing of space. *Phenomenology & Practice, 10*(1), 25–46.

McCormack, D. (2010). Thinking in transition: The affirmative refrain of experience/experiment. In B. Anderson & P. Harrison (Eds.), *Taking-place: Non-representational theories and geography* (pp. 201–220). Abingdon, United Kingdom: Routledge.

Merleau-Ponty, M. (2005). *Phenomenology of perception.* Abingdon, United Kingdom: Routledge.

Miller, D. (Ed.). (2001). *Car cultures.* Oxford, United Kingdom: Berg.

Moran, D. (2000). *Introduction to phenomenology.* Abingdon, United Kingdom: Routledge.

Nail, T. (2019). *Being and motion.* Oxford, United Kingdom: Oxford University Press.

Pink, S. (2012). *Doing sensory ethnography.* Newcastle upon Tyne, United Kingdom: Sage.

Raingruber, B. (2003). Video-cued narrative reflection: A research approach for articulating tacit, relational, and embodied understandings. *Qualitative Health Research, 13*(8), 1155–1169.

Rice, D. H. (1989). Whitehead and existential phenomenology: Is a synthesis possible?. *Philosophy Today*, *33*(2), 183–192.

Robert, F. (2005). Whitehead and phenomenology: An intersecting reader of the Late Merleau-Ponty and the Whitehead of process and reality. *Chiasmi International*, *7*, 366–367.

Robinson, K. (2005). Towards a metaphysics of complexity. *Interchange*, *36*(1–2), 159–177.

Shade, P. (1995). *Ron L. Cooper, "Heidegger and Whitehead: A phenomenological examination into the intelligibility of experience" (Book review)* (Vol. 31, pp. 246–253). Buffalo, NY: The Society.

Sheller, M. (2005). Automotive emotions: Feeling the car. In M. Featherstone, N. Thrift & J. Urry (Eds.). *Automobilities* (pp. 221–242). Newcastle upon Tyne, United Kingdom: Sage.

Sheller, M. (2014). The new mobilities paradigm for a live sociology. *Current Sociology*, *62*(6), 789–811.

Sheller, M., & Urry, J. (2000). The City and the car. *International Journal of Urban and Regional Research*, *24*(4), 737–757.

Sheller, M., & Urry, J. (2006). The new mobilities paradigm. *Environment and Planning A*, *38*(2), 207–226.

Sherburne, D. W. (1983). Whitehead, categories, and the completion of the Copernican revolution, *The Monist*, *66*(3), 367–386.

Stengers, I. (2011). *Thinking with Whitehead: A free and wild creation of concepts*. Cambridge, MA: Harvard University Press.

Tarde, G. (2000). Social laws: An outline of sociology, trans. Kitchener, ON: Batoche Books.

Taylor, N. (2003). The aesthetic experience of traffic in the modern city. *Urban Studies*, *40*(8), 1609–1625.

Thrift, N. (2004). Driving in the city. *Theory, Culture & Society*, *21*(4–5), 41–59.

Thrift, N. (2008). *Non-representational theory: Space, politics, affect*. Abingdon, United Kingdom: Routledge.

Urry, J. (2004). The "system" of automobility. *Theory, Culture & Society*, *21*(4–5), 25–39.

Urry, J. (2005). The complexity turn. *Theory, Culture & Society, 22*(5), 1–14.

Urry, J. (2010). Mobile sociology. *The British Journal of Sociology*, *61*(s1), 347–366.

Walsham, G. (1997). Actor-network theory and IS research: Current status and future prospects. In A. S. Lee & J. Liebenau (Eds.), *Information systems and qualitative research* (pp. 466–480). Boston, MA: Springer.

Weber, T. (2016). Metaphysics of the common world: Whitehead, Latour, and the modes of existence. *The Journal of Speculative Philosophy*, *30*(4), 515–533.

Whitehead, A. N. (1978). *Process and reality: An essay in cosmology (corrected edition)*. New York, NY: The Free Press. Originally published in 1929 by Macmillan.

Whitehead, A. N. (1985). *Symbolism, its meaning and effect*. New York, NY: Fordham University Press, The Macmillan Company.

3 History, capitalism, and the ethics of kinetics

The novel *The Go-Between* by L. P. Hartley (1954) begins with the sentence, "The past is a foreign country: they do things differently there" (p. 3). This famous opening sentence suggests the past as a somewhat out-of-reach exotic locale, the keys to which are old relics drenched in nostalgia, found, in this case, in a "rather battered" old box (p. 3). When I first came to live in Ho Chi Minh City (HCMC), coffee was served through iconic steel drip filters called *caphe phin*, whose provenance is with the French, but in true Vietnamese style, have since been modified in design and usage. Even then, 10 years ago, these devices seemed at odds with the pace of the world, a relic from another time and place. Upon entering a café, one is presented with a coffee cup with some sweet milk in the bottom and a steel coffee filter perched on top containing hot water and coffee. One is then required to sit patiently (like waiting for a webpage to load on a slow Internet connection) as the thick coffee slowly drips into the cup. This is slow coffee, more at home in the 1950s Saigon of Graham Greene's *The Quiet American*, where women wearing the traditional dress of *áo dài* elegantly strolled past French colonial buildings, across wide-open boulevards, and where only the sounds of bicycles and *cyclo* riders competed with the gossip and conversation of the *flâneurs*. In the days when HCMC was called Saigon, it was a city where only the Saigonese, who saw themselves as an educated people with international influences, could legally live and work. In contrast, the whole city now seems up for grabs, as people arrive from all provinces to find opportunities, earn money, buy land and houses, and attend universities, often staying on in HCMC once they graduate.

An important dimension of the mobilities that course through a city is the dimension of rhythm. Rhythm is repetition and comparison of time and space, how close behind one thing follows another (Edensor, 2010; Lefebvre, 2007). Juxtaposed rhythms that seem at odds with each other, conflicting or dissonant, what Lefebvre refers to as "arrhythmia" (2007, p. 16), seem more difficult to sustain over long periods of time. These *caphe phin* filters, at odds with the speeds of contemporary HCMC, have become something of a rarity in the city's cafés, now mostly only found in the traditional Vietnamese-style cafés or used as a super-sized industrial form in the coffee franchise chains, dispensing much larger quantities of coffee (more drips in less time). That these filters have not disappeared altogether is either testament to the nostalgic vein that runs through Vietnamese culture or because they work best with Vietnamese coffee.

Throughout the years, as the *caphe phin* coffee filters dispensed both coffee and time in quiet meditative drips, Vietnam and its people experienced decades of instability, extreme hardship, war, famine, and political chaos and upheaval. Though Vietnam is now considered a lower-middle-income country (Hansen, 2015), with a fast-growing economy, only four decades ago, it was one of the poorest countries in Asia (Hansen, 2015), and even as recently as the 1990s, Vietnam had more poverty than any large South East Asian nation, with half of the rural population not getting the recommended minimum requirements of calories (Elliot, 1995; Kolko, 1997). Additionally, during the period from 1944 to 1945, a famine caused somewhere between 400,000 and 2,000,000 deaths, and this was then followed by a series of wars, from 1945 to 1975 (Barbieri, Allman, Pham, & Nguyen, 1996).

Vietnam has been influenced by many different colonial influences including the USSR, China, the United States, and the French, whose 70-year rule was only overthrown as the result of a war in 1954, which preceded another 20 years of war known as the 'Vietnam' or 'American War,' which officially ended with the fall of Saigon in 1975. From 1975 onwards, Vietnam continued to face extraordinarily difficult economic times, which prompted large-scale economic reforms in 1986 known as *đổi mới*, whereby the country transitioned from a socialist, centralized economy to a market-led one, which heralded a trajectory of economic development that continues today.

This history of influences has left Vietnam with a potpourri of different models of governance, laws, regulations, and regimes of logic that often result in complex and sometimes contradictory overlapping structures. The result is that Vietnamese social structures, including formal and informal linguistic structures, institutional structures such as banking and the taxation system, and traffic structures can be extremely complex, obtuse, and vague. Navigating such bureaucratic complexities often requires patience, and advanced skills and knowledge in the ways of doing things in Vietnam and a somewhat adventurous and entrepreneurial sensibility. These institutional structures can sometimes be so obtuse and complex, in fact, that it is only through fluid practices that circumvent the 'rules' that anything could proceed at all. What appears as rigid, therefore, can also paradoxically open up new kinds of timespaces, where new qualities and new fluid dynamics can emerge (Thrift, 2008). This complex structuration in Vietnamese society has a twofold character, in that, whilst the lack of transparency and complexities might cause confusion or lower efficiency on one hand, it also opens up possibilities for fluidity in the systems, paradoxically, often *increasing* efficiency in different ways. Of course, such outcomes are encouraged if accompanied by sensibilities and procedures characterized by flexibility, open to transcending perceived boundaries, and by recourse to multiple levels of authorities. For example, funeral parties (which, in HCMC, continue without pause for three days and three nights – with a live band) often erect marquees that extend from the house out onto the street. Whilst this often disrupts traffic or even completely blocks necessary and busy traffic thoroughfares for three days, it is easily achieved and considered a normalized and accepted part of paying respects to the deceased.

These kinds of tailor-made infrastructural developments are quite commonplace in HCMC, and citizens are generally familiar with these ways of working. For example, local residents often mix concrete to fill potholes in the road (see Figure 3.1) or build small access structures such as motorcycle ramps onto footpaths, and even sometimes erect their own versions of streetlights.

Such practices straddle boundaries of the formal and the informal, as local residents are sometimes paid to act as caretakers of local smaller streets, whose duties might involve keeping an eye on things or carrying out small-scale maintenance work, such as clearing the streets and drains of leaves and rubbish. Though the traffic in HCMC can be a little dog-eat-dog, as space is very limited, should an unfortunate event occur, such as a traffic accident, people are quick to lend a hand. Especially amongst the users whose daily working space is within the traffic environment, such as motorcycle delivery riders, there is a noticeable and growing collegiality. There seems a tacit agreement that these are difficult working conditions that require support, and such users are often seen helping others in distress or resolving problems, such as when an overhead cable unravels and drapes itself across the road.

HCMC is the business city in a country that only truly embraced capitalism in the last 30 years, and while the Vietnamese capital, Hanoi,

Figure 3.1 Local residents fill potholes with concrete. Copyright 2017 by C. Tran.

located in the north, is a more tightly controlled city, HCMC, as the economic powerhouse of the country, is given more free rein. The current construction boom is driven by disconnected and often incongruous projects occurring within structures and processes, whose checks and balances can be inconsistent and beyond the control of any single determining body. During the last six or seven years, in particular, there has been a profound *verticalization* process in HCMC that seems almost out of control. Schwenkel (2013), in her paper on post-socialist affect in Vietnam, describes how the vision of urban futurity demands not only a radical break from the past but also aesthetic forms of *high* modernity, where 'high' literally means tall buildings that signify wealth and technological advancement. These rampant construction projects also inject a feeling of urgency into the traffic system, as vehicles such as concrete trucks race all over the city, pressured by tight timelines, feeding the seemingly insatiable needs of the construction industry. Many new high-rise apartment buildings are accessed via very small streets and the additional numbers of cars and motorcycles created by newly filled apartment complexes can be problematic on the streets below, often compounded by other factors such as flooding.

For those in a position to make substantial economic gains from this seemingly exponential surge of activity, there is a lot of money to be made, as evidenced by the number of Rolls Royce, Maserati and all manner of expensive cars on the streets, and while there is a growing middle class, the widening economic gap between the rich and the poor is also evident. Whilst salary is, by no means, an indicator of actual income in Vietnam, the salaries of the lower paid occupations is noteworthy, such as factory workers and security guards, who represent a quite substantial sector of the population, and whose salaries remain extremely low for a city that is increasingly expensive for basic items. For example, according to a 2016 Living Wage Report (Research Centre for Employment Relations, 2017), salaries for garment workers were estimated to be 259 USD per month.

The complex structuration that exists in Vietnam, as well as the processes by which major projects are funded, impacts infrastructure development in ways that are often impossible to predict. For example, the 19-kilometre metro line construction project, currently underway in HCMC, which includes a subway system and skytrain, was, in October 2017, reported to be at risk of ceasing construction, because the Japanese company contracted for the project was not receiving payments. Le Nguyen Minh Quang, head of the management board for the urban railway project, known as the Saigon Metro, said that the problem was "not lack of money, but of a proper mechanism for capital disbursement" (Tuoi Tre News, 2017, para. 5).

Large infrastructure projects, especially those involving international companies and institutions, often encounter similar challenges. The Bus Rapid Transport (BRT) project, a large-scale project in Hanoi and HCMC begun by the World Bank, is a case in point. Kim (2017) explains that the competitive free-for-all environment meant that other players entered the project after all the groundwork and plans had been finalized. The result of this was that additional players who entered the deal later, were able to slice out important pieces of the project for themselves, for example by taking ownership of the high-ridership lines necessary for providing the revenue needed to recoup the initial investment. These changes modified the original vision of the project in ways that diluted its effective implementation, introducing new aims, and so what began as a single cohesive design was then transformed into a multiplicity of aims and agendas. According to Kim (2017), once the BRT was completed in Hanoi, the bus lanes, intended for the exclusive use of BRT buses, were later opened up to government buses and private cars, resulting in functional flaws and inefficiencies in the system.

The notion of *autopoiesis* is a central concept in complex adaptive systems theory, and though there exist different and competing ideas concerning the implications and employment of the notion (Byrne & Callaghan, 2014), it basically refers to the ability of a system through the existence of itself – its processes and structures – to (re)produce itself (Zeleny & Hufford, 1991). In other words, an autopoietic system is one that self-organizes and self-creates, as it defines, maintains, and reproduces itself in its own image, with a cohesive or persistent identity (Fuchs & Hofkirchner, 2010). Urry (2004) says that an autopoietic system generates "the preconditions for its own self-expansion" (p. 27). Urban traffic systems are said to be autopoietic, and Urry has commented that, "automobility can be conceptualized as a self-organizing autopoietic, non-linear system that spreads world-wide" (2004, p. 27). Whilst the traffic system in HCMC may be autopoietic, evolving a consistent aesthetic and reproducing itself in its own image, according to its own needs and goals, it is, unsurprisingly, a disparate, multiplistic, and fragmented collectivist project.

The constitution of the kinetic style of traffic in HCMC emerges in conjunction and in response to the very being of the traffic, which is to flow, no matter what. Practices, ethics, movement, and rhythms that synergize with this aim tend to endure, and the amalgamation of the multitudes of individual movements by traffic users is significantly and fundamentally influential on the evolution of the system. As a largely self-creating system, whose principles of organization reflect and conform to historical – and sometimes chaotic – structures, a potpourri of often contradictory logics that evolved through various colonial influences, and the ethics of a kinetically obsessed modern existence, this system also evolves through the spontaneous actions and reactions of individual traffic users coping as best they can within a somewhat manic, pressure-cooker environment; in other words, it's complex. Coupled with that, the traffic is growing at an uncontrollable rate, linked to the exponential migration of people from the rural areas to the city, the increasing numbers of people who now own motorcycles and cars due to the rise in incomes, and the abolition of import taxes on some cars.

The outcome of all of these processes and influences is the emergence of a particular aesthetic, order, character, and style, that somehow, like the 'glue' of socialist common purpose in a former era (Elliot, 1995), holds the whole thing together.

In reference to the marginal status of residents in informal slum settlements in Zanzibar, who are denied formal rights to their places of habitation, Bissell (2011) points out:

> The city is a precipitate shaped by history, and what remains or endures is *hardly accidental* [emphasis added]. In other words, place – and what stays in place – is always linked to social processes and broader questions of power. (p. 7)

In a sense, everything that is, is not really accidental, because structures were in place that resulted in just that outcome. Though a novel outcome, it is hardly accidental when motorcyclists find themselves blocked by cars that have taken over the motorcycle lanes and are then forced to mount the sidewalk and ride across an empty field (which, over time, becomes a track, which is later formalized and asphalted by the roads and transport department to become a permanent part of the traffic infrastructure), as can be seen in the line of motorcycles pictured (see Figure 3.3). The motorcycles shown in the photograph are not traveling on an existing road. In fact, just a few days before this picture was taken, the route the motorcycles are seen traversing was not even any kind of a path. This photograph was taken from the road where these motorcycles *should* actually be, but the usual road was impassable for motorcycles due to the large numbers of cars that had taken over all lanes, including the lane reserved for the motorcycles.

Figure 3.2 Fluid infrastructure. Copyright 2017 by G. Wyatt.

Figure 3.3 Car traffic jams force motorbikes to forge their own paths through vacant land. Copyright 2017 by C. Tran.

The organization of the HCMC traffic system evolves through each entity reacting to the limited information around it, engaged in *absorbed coping*. Every day, individual traffic users in HCMC, through diverse creative actions, forge new trajectories, leaving their marks and instigating novel changes in topology and topography. Sometimes, the traffic authorities act once the traffic has already forged its own path, formalizing the informal roads made by traffic users, letting the traffic, at least to some extent, decide just how it will flow. New pathways or practices may then become enduring elements in the system over time, either through the mass consensus of users or through more formal endorsement by the government, such as by way of concrete infrastructure building or the formation of new laws. This can result in concrete infrastructure and networks of practices that reflect and are, in fact, solidified forms and manifestations of the dynamics of the traffic itself, objectifications of deep modes of existence, paradigms, logics, and ways of being. Of course, the process is co-constitutional: individual traffic entities respond to the environment, with actions that create enduring forms (infrastructures or patterns), which users then respond to, which then create more enduring patterns and forms, and so on.

On my previous route to work, I daily rode my motorcycle through a section of road that was always heavily congested during peak hour traffic time, causing many motorcycle riders to mount the curb and travel along the pedestrian sidewalk in order to make a right turn a little further down the path. This made for an extremely disorderly situation, slowly destroying the paved sidewalk, and creating a dangerous situation for pedestrians. After some months of this, the transport authorities responded by digging up what was left of the paved sidewalk and creating a dedicated asphalt right-turn lane for cars and motorcycles. In other words, a formal road was created that followed the trajectory of the informal path made by motorcyclists, born from coping strategies, whereby individual users simply *overflowed* beyond boundaries. Now, even years later, this new formal turning lane still exists, but, because its design followed the organic inclinations of the traffic flow, the curvature of the turning lane is quite sudden, and it also leaves vehicles with two possible points at which to turn, earlier or later, which can create confusion for riders and drivers. Also, despite these changes, at busy traffic times, the motorcyclists still mount the curb and ride along a very narrow sidewalk that has been left just next to the dedicated turning lane in order to avoid the jam and confusion.

References

Barbieri, M., Allman, J., Pham, B. S., & Nguyen, M. T. (1996). Demographic trends in Vietnam. *Population: English Selection, 8*, 209–234.

Bissell, W. C. (2011). *Urban design, chaos, and colonial power in Zanzibar.* Bloomington, IN: Indiana University Press.

Byrne, D., & Callaghan, G. (2014). *Complexity theory and the social sciences: The state of the art.* Abingdon, United Kingdom: Routledge.

Edensor, T. (2010). Introduction: Thinking about rhythm and space. In T. Edensor (Ed.), *Geographies of rhythm: Nature, place, mobilities and bodies* (pp. 1–18). Farnham, United Kingdom: Ashgate Publishing, Ltd.

Elliott, D. W. (1995). Vietnam faces the future. *Current History, 94*(596), 412.

Fuchs, C., & Hofkirchner, W. (2010). Autopoiesis and critical social systems theory. In R. Magalhães & R. Sanchez (Eds.), *Advanced series in management* (Vol. 6, pp. 111–129). Bingley, United Kingdom: Emerald Group Publishing Limited.

Hansen, A. (2015). The best of both worlds? The power and pitfalls of Vietnam's development model. In A. Hansen & U. Wethal (Eds.), *Emerging economies and challenges to sustainability* (pp. 108–121). London, United Kingdom: Routledge.

Hartley, L. P. (1954). *The go-between.* New York, NY: Alfred A. Knopf.

Kim, H. (2017, September 22). *Reassembling transportation infrastructure in Ho Chi Minh City.* Public lecture. Ho Chi Minh City, Vietnam: An Ordinary City Series at Coeverything.

Kolko, G. (1997). *Vietnam: Anatomy of peace.* London, United Kingdom: Routledge.

Lefebvre, H. (2007). *Rhythmanalysis: Space, time and everyday life.* New York, NY: Continuum.

Research Centre for Employment Relations, Global Living Wage Coalition. (2017). *Living wage report: Urban Vietnam: Ho Chi Minh City: with focus on the garment industry,* 2016. (Report No. 10). Retrieved from https://www.isealalliance.org/sites/default/files/resource/2017-12/Urban_Vietnam_Living_Wage_Benchmark_Report.pdf

Schwenkel, C. (2013). Post/socialist affect: Ruination and reconstruction of the nation in urban Vietnam. *Cultural Anthropology, 28*(2), 252–277.

Thrift, N. (2008). *Non-representational theory: Space, politics, affect.* New York, NY: Routledge.

Tuoi Tre News. (2017, October 14). Official warns of 'unimaginable consequences' if Saigon metro ceases construction. *Tuoi Tre News*. Retrieved from
https://tuoitrenews.vn/news/society/20171014/official-warns-of-unimaginable-
consequences-if-saigon-metro-ceases-construction/42045.html

Urry, J. (2004). The 'system' of automobility. *Theory, Culture & Society*, *21*(4–5),
25–39.

Zeleny, M., & Hufford, K. D. (1991). All autopoietic systems must be social systems
(living implies autopoietic. But, autopoietic does not imply living): An application of autopoietic criteria in systems analysis. *Journal of Social and Biological
Structures*, *14*(3), 311–332.

4   The bifurcation of nature
Matters of fact and matters of concern

Whitehead's process philosophy, also called the philosophy of organism, is a reaction to historical developments that led to what he saw as the demarcation of nature into two distinct realities or realms: on one side, that of scientific constructions and relationalities, built upon the universal objectifications of knowledge – in other words, 'facts' – and, on the other, the more value-laden world of experience and subjectivity, the realm of emotions, feelings, and intensities. The prioritization of objective facts, according to Whitehead, creates the situation whereby the value-laden relationalities of experiential complexity and subjective perception come to be viewed as fanciful additions *to* nature, rather than things that actually exist *in* nature, equal to – even if different from – scientific 'facts'; a situation, Whitehead refers to as *the bifurcation of nature*. Rather than denounce the scientific viewpoint, Whitehead – being a man of science and a mathematician as well as a metaphysician – suggests the solution lay in the acceptance of both versions of reality, each being valid, but within different contexts. This approach, he sums up in the oft-quoted sentence, "For natural philosophy everything perceived is in nature. We may not pick and choose" (Whitehead, 1920, p. 29).

To admit the abstractions of science as, the 'real' stuff, the fundamental building blocks of nature, but to view abstractions arising out of experiential complexity as fanciful additions of the mind, is to settle for an incoherent and incomplete account of nature. As Whitehead (1920) explains:

> You cannot cling to the idea that we have two sets of experiences of nature, one of primary qualities which belong to the objects perceived, and one of secondary qualities which are the products of our mental excitements. All we know of nature is in the same boat, to sink or swim together. (p. 148)

The answer to this bifurcation is not to choose to side with one team or another. Latour (2004) refers to Whitehead's bifurcation as the divide between, what he calls *matters of fact* and *matters of concern*, and says that though matters of fact are "a poor *proxy* (orig emphasis) of experience" (p. 245), the solution is not to attack, criticize, or historicize them as "made up," or "interpreted" (p. 245), nor to expose them as "a confusing bundle of polemics, of epistemology, [and] of modernist politics" (p. 245). Neither, according to Latour, is the solution to "flee" (p. 245) back into mind and add to facts the cultural and symbolic dimensions. Instead, Latour (2004) opts for a third path in the development of a realist attitude that draws on Heidegger's (2001) notion of the *gathering*, a productive tool in viewing *blackboxed* objects (settled matters of fact) in all their complex relationalities. The term 'traffic,' itself, is an enormously compressed backboxed abstraction, one of those universal referents that Latour (2005) describes as performing a "sleight-of-hand" that substitutes a complex, multiple, and specific performance of fluid and contingent relations, for a static, simple, universal catch-all term, in such a way that the latter might explain the former (p. 100). The entity 'traffic,' as a universal abstract object that exists in the world, can never be used to explain contingent concrete performances with unique spatial and temporal characteristics. Yet, as we shall see in the following chapter, certain research approaches to 'traffic,' such as computer simulations – the purpose of which, surely, is to better understand real-world phenomena – have come to be increasingly divorced from actual, concrete, place-based traffic events, and are often constructed using universal notions that can be transferred across different kinds of systems, not only traffic systems.

Whitehead (1920) says, "for us, the red glow of the sunset should be as much part of nature as are the molecules and electric waves by which men (sic) of science would explain the phenomenon" (p. 29). Though our subjective experience of *redness* in a sunset-event may also be experienced as feelings of psychological warmth – romance or nostalgia, for example – these 'additional' ingredients also constitute the event and should not, therefore, be considered less valid than the scientific elements of molecules and electric waves. It is for this reason that Whitehead begins with the 'event' as the fundamental building block of the universe. The event not only allows for those meaningful and experiential influences and dimensions of existence but can also show how nature is built up from these more subjective dimensions of perception.

The danger in this 'bifurcation,' is when complex and particular, value-laden, concrete actual performances come to be substituted or displaced by simple, independent, and universal matters of fact, giving rise to the paradoxical situation, as Latour says, in being given more, we end up settling for less (Latour, 2008). In other words, the contingent relations of the actual concrete events that gave rise to the universal 'law' are substituted by independent entities we call 'facts,' so that the referent or actual performance comes to be displaced by its reference, then assuming the status of a reified causal explanation. Latour (2014) has described how scientific processes create universal matters of fact by taking the actual actions and "performances" of matter (p. 3) and separate them of all local contingencies, associated with the "who," the "what," and, especially, the "when." Through their empirical studies on research processes in scientific laboratories, Latour and Woolgar

(1986) were able to locate the point in scientific processes when something that began simply as a statement became a fact, "freed from the circumstances of its production" (p. 105). Law and Mol (2001) write that once a fact was established in a single place, it was "supposed to transport itself everywhere, free of cost and without effort" (p. 609).

For example, Latour describes how the mobility of the abstract technical object, the cartographic map, opened corridors of common languages that enabled mobility through immutability. Maps, as matter-of-fact objects, are "immutable, presentable, readable and combinable with one another" (Latour, 1986, p. 7) and so able to keep their relations intact whilst being mobilized in order to muster allies and convince others of certain states of affairs. A map is an example of what Latour calls an *immutable mobile*, a notion that forms an important foundation in the development of actor-network theory (ANT). A map is also a blackbox, an abstract object that makes opaque the complex *gatherings* and enfoldments of time and space and the processes and performances that occur behind the scenes and actually go into making a complex thing what it is (Latour, 1999). The example of cartographic maps shows how common corridors and infrastructures of *technicity*, constituted of shared paradigms and languages, not only allow the transportation of fact-objects across time and space, but in doing so, also constitute temporalities and spatialities. Matter-of-fact abstractions, such as maps, emerge from attempts to objectify and mobilize knowledge by rationally defining and measuring reality in universal terms, creating distinctions that attempt to categorize, disentangle, and make independent phenomena that, in reality, are complexly entangled. In this way, matter-of-fact processes decontextualize knowledge from its places of production by cutting meaningful relations.

From this seemingly simple abstract artefact, the mobile cartographic map, fact-friendly infrastructures embedded in certain paradigms, what Simondon (2011) might call infrastructures of *technicity*, now take up much more space in the world, constituting a kind of post-phenomenological *ethnomathematical* moving frame of spatiality and temporality (Thrift, 2008). This displacement of our more complex and value-laden realities, *matters of concern*, with *matters of fact*, has profound effects on our experiences and awarenesses in the world, but also on our understandings of the world and how we see ourselves and our place in it.

According to Whitehead, the bifurcation of nature occurred as the result of two interrelated historical developments:

- The process of the 'mathematization' of nature by natural philosophers and scientists (Weber, 2016, p. 518), by way of an increasing reliance on mathematics (the form of which became increasingly abstract) as a language for rendering and explaining nature, and previous to this,
- The extension of the view of the world as composed of two different classes of things: substance and qualities or subjects and predicates.

When nature is reduced to such entities as universal laws, molecules, waves, and mathematics, we end up with a situation where "there is no light or colour as a fact in nature" (Whitehead, 1948, pp. 54–55), as these kinds of emergent experiential phenomena cease to exist when reduced to such fundamental elements. McIntyre (2010) reminds us that "the smell of something has no correlate at the microphysical level" (p. 5), so, there is nothing at the primary level of description in chemistry that can recover what it means for a chemical compound to have a particular smell. At this reductionist level of chemistry, smell has been stripped of its meaningful dimensions, and though it still exists in the world, it is a product of very different relations. For this reason, Stengers (2011) describes this version of nature as "colorless, odourless, and mute" (p. 74).

As a way of understanding how relationalities of matters of fact and matters of concern produce abstract objects that are both 'real,' but real in different senses, we can, as an example, see how the entity or object 'red' arises from very different relations, existing, as it were, in different 'worlds' or contexts. The object 'red' exists, in one incarnation, as a technical object, such as the 'red' of a hex triplet for HTML applications (#FF0000), and also the expression of red in a sRGB colour space (255, 0, 0). Also, 'red' exists as a scientific abstraction, namely, the molecules and electric waves by which science explains the existence of 'red' (Whitehead, 1920, p. 29), and, then again in a more subjective way, as "the red glow of the sunset" – a particular subjective experience of redness. Whitehead (1920) would call the latter, a "percipient event" (p. 107), where 'redness' can never exist independently, but only ever as a term in a multiple relation. In admitting only certain versions of 'red' into nature, however, and dismissing the others as invented psychological additions, as fanciful productions of mind and therefore not *in* nature, we are creating a bifurcated view of reality. Whitehead's answer to this is that all of these 'reds' are 'real,' but they are "real in different senses" (1920, p. 30).

How did the situation arise, whereby one description of 'red' has come to assume the status of 'real' over others? According to Whitehead, the bifurcation of nature has its roots in the ancient Greek subject–predicate model, which began as a linguistic form (for example, 'that stone is grey') but later expanded and generalized to include all of reality. As a generalized scheme, this model divided the world into primary substances qualified by universal qualities: matter (stones), known to us only through sensory perception of their attributes (such as greyness) (Whitehead, 1978). This classical view of reality maintains that while attributes or qualities may only be known through perception, there always remains (*in* nature) an enduring 'something' beyond our awareness: 'substance' or 'matter.' In this way, substances and qualities came to form two different classes of reality (Hosinski, 1993), and all objects become known in terms of subjects and predicates, substance and qualities, or particulars and universals (Whitehead, 1978). A 'particular' then becomes a substance that is qualified by some quality, and a 'universal' becomes a quality that is able to qualify many different substances (Hosinski, 1993). Due to the principle inherent in the subject–predicate model: that substance can only be known through our awareness of qualities and through sensorial experience, knowing the world through sense-perception then became the dominant foundation and method for developing a theory of knowledge of the world (Hosinski, 1993).

Whitehead (1920) says that later developments in the 17th century then "completely destroyed the simplicity of this 'substance and attribute' theory of perception" (p. 27). This occurred through the systemization of the transmission theories of light and sound, notably, the connections between light and colour, as posited by Newton. With Newton's physics of light, our experience of 'greyness' came to be viewed as the result of the transmission of light waves and, therefore, could no longer be viewed as an attribute of the substance itself. Such an attribute could no longer, therefore, belong to the object as a primary quality, and this meant that it must now be an addition of mind: a secondary quality that does not have its origin in nature.

From here, two things occurred, whose implications still continue: a split was created between the subject and object – the 'Cartesian divide' – whereby 'actual' things could be independent, requiring nothing but themselves in order to exist, and, second, subjects having conscious experiences (the interior world of mind) became viewed as the most appropriate source of data for all philosophical enquiry (Whitehead, 1978). The abstract notion of substance, which was earlier, a productive *conceptual* tool in philosophy, was then imported into science as concrete 'matter' in the form of an actual enduring substance in nature that underlies our perceptual experiences with attributes and qualities. This enduring 'real,' yet abstract, thing we call 'matter,' now viewed as actually existing in the world, might remain meaningful for science, but has no real meaningful existence insofar as our value-laden, experiential lives are concerned, hence, the bifurcation.

The creation of universal laws and matters of fact that are able to be transferred across different kinds of systems as well as the creation of infrastructures that facilitate the mobility of such objects then became the dominant pursuit of scientific endeavour from the 17th century onwards. Towards these ends, science needed tools useful for the abstracting of nature that were characterized by neutrality, and according to Whitehead (1948), these qualities were to be found in a new kind of mathematics that differed in nature from that of earlier epochs. The role of mathematics is, therefore, so integral to the epistemological activities of classical science, that Whitehead (1948) has suggested that it was mathematics that "supplied the background of imaginative thought with which the men of science approached the observation of nature" (p. 32). As mathematics "withdrew" (Whitehead, 1948, p. 34) into more extreme abstraction, by becoming separated from the particular sets of entities that it was used to explain, it simultaneously became more important and more ubiquitous in the analysis of the concrete fact. In this way, the universal came to displace the particular, even to the degree that such laws come to be endowed with independent and reified causal power. The appeal and productiveness of the qualities of mathematical abstraction then opened the way for algebra, itself, composed of abstractions that "refer impartially to any number" (Whitehead, 1948, p. 31), which then led to the proliferation of algebraic algorithms that now increasingly constitute the backdrop of modern existence.

This 'solid' bedrock of facts and universal laws has been built up, historically, through processes of classical empiricism, which are not adept at dealing with other forms of awarenesses, such as those more embodied, pre-intellectual, and vague experiences that Whitehead calls *causal efficacy*. In other words, classical scientific practice is based upon the principle of *omission* by dismissing and neglecting those modes of experience deemed irrelevant, whilst allowing modes of thought that are often less intuitive (Whitehead, 1968, p. 74). As Latour (2008) points out, this classical perspective of the real, as being constituted of molecules and light waves, is, simultaneously, a view from everywhere and also a view from nowhere. Molecules and light waves are 'meaningless' so far as our meaningful perception of them in experience is concerned, and this kind of reductionism, according to Whitehead (1948), makes nature "a dull affair, soundless, scentless, colorless; merely the hurrying of material, endlessly, meaninglessly" (p. 56). As Latour (2011) has decreed, there should not be the situation, whereby certain modes of relationalities are viewed as more valid than others and where one discipline, such as modern mathematical physics, should come to be viewed as the "final arbiter" of another (p. 5).

To view traffic from 'nowhere' is to conceive of it as 'merely the hurrying of material, endlessly and meaninglessly,' and yet, we know that traffic is a phenomenon drenched in – indeed, constituted from – meaningful relations and subjective felt intensities, emotions, and vague, embodied awarenesses. To consider traffic as merely an exercise in a kind of geometry-in-action, a collection of roundabouts, T-junctions, right angles, crossroad intersections, and measurable matter locatable by mathematical coordinates moving at a measurable mathematical velocity is to miss the essence of the phenomenon and to bifurcate nature. To supplant the complex experiential relationality of traffic with *mathematical* relationality is to fall prey to what Whitehead (1948) calls *the fallacy of misplaced concreteness*, which he describes simply, as, "mistaking the abstract for the concrete" (p. 52) by "neglecting the degree of abstraction involved when an actual entity is considered merely so far as it exemplifies certain categories of thought" (Whitehead, 1978, p. 7).

Some approaches to traffic research, such as computer traffic modelling and simulations, that attempt to quantify experience and objectify knowledge, are embedded in paradigms of technicity, and create scientific abstractions and technical objects that maintain an inherent Cartesian duality between subject and object. One of the (many) dangers in this is the attraction towards *technological solutionism*, devoid of philosophical contemplation, as the way forward in the solving of complex modern problems, sometimes called 'wicked problems.' Heidegger (1977) has famously said, "the essence of technology is nothing technological" (p. 35), and while "we represent technology as an instrument, we remain held fast in the will to master it." (p. 32). One problem with research that views traffic as constructed of technological objects or as a phenomenon that can be controlled or "solved" through technological means is that the whole endeavour can begin to form a loop whereby the *means* of the research become so entangled with the *goals* that the means and goals become inseparable. The methods by which we gather the raw data then may implicitly influence the kinds of data collected as well as the interpretation of data. The danger in this is that technology becomes the means and the ends, setting the agenda and influencing our values, goals, and how the problem may be approached or framed. As Dreyfus (2007) has said, the problem with modern mathematical natural science as formulated by Descartes, Newton, and others is not its paradigm of exactness, but that beings are only able to be discovered in the only way that they can. For an entity to be able to exist at all, it needs to fit into the criteria of the model of representation and the particular nature of the relations and parameters by which the phenomenon is being framed and articulated. In other words, as Dreyfus (2007) points out, Descartes decided for us all what counts as matter, what kinds of proof are needed, and what methods may be used in discovering it.

The problem of the bifurcation does not lie with abstractions themselves (abstractions are necessary for any meaning to emerge in the world at all), but with changes in modes of awarenesses, aesthetics, and logics that, when proliferating en masse, conceal, and unconceal worlds. Abstractions that emerge from meaningful experiential complexity are different in nature from those emerging from the processes of classical science and from infrastructures of technicity, because they are constituted by different kinds of relations. In a computer simulation, for example, the structures of experiential complexity are impossible to replicate due to their nonlinear nature, the involvement of the vague feelings of *causal efficacy* in processes of meaning-making, and the sheer number of quasi causes or influences involved in the event. Individual subjective experience is often based on particularities, contingent relationships that lack stability and universalism, and often includes dimensions of felt intuitions and synaesthetic embodied and unconscious processes. These complex factors mean that it lacks objectivity, is, in fact, not easily objectified, not easily transferable across different kinds of systems, and, therefore, also not technically efficient. Whitehead (1978) points out: "There are no brute, self-contained matters of fact" (p. 14) and argues that any attempt to analyse and express a moment of experience would necessarily lead us beyond the matter of immediate experience and into the past and/or the future and to the eternal realm of pure possibilities.

A factual entity existing independently, without recourse to other places and times, requires a classical Aristotelian view of time and space as a kind of room – what Sloterdijk (2012) calls "container-physics" (p. 37) – within which, sits matter, or as Whitehead (1920) puts it, "the unconscious presupposition of space and time as being that within which nature is set" (p. 20). Such a view requires time and space to be independent and constant, and yet, we all have probably had experiences in traffic that suggest time and space are not constant. For example, when a trip that seems short when measured in kilometres seems to take much longer than expected, or conversely, when our destination suddenly pops out in front of us, as though from out of nowhere. The fact that time and space can compress or extend, depending on experience, and are neither constant, nor independent from each other, is an inconvenient truth for sciences that value constancy, efficiency, and the objectification of knowledge. Despite this experiential reality, whole infrastructures of traffic and tools for traffic research are embedded in frameworks of representation that require time and space to not only be independent from each other, but to be constant, so that an entity is able to exist in what Whitehead calls "simple location."

The *fallacy* (yes, it's another fallacy) *of simple location* as described by Whitehead (1948) is when, "Material can be said to be *here* in space and *here* in time, or *here* [original emphasis] in space-time, in a perfectly definite sense which does not require for its explanation any reference to other regions of space-time" (p. 50).

Simple location assumes that there may be fixed points in a uniform space and time, such as in a Cartesian and Euclidean space, whereby matter can be said to have the property of being in a mathematical point of space and time that has no reference to other *meaningful* regions of space and time. In other words, simple location ignores the meaningful ways in which entities come to be just what they are, constituted with value and significance, and this notion, together with the fallacy of misplaced concreteness, is how we arrive at the bifurcation of nature. The problem is then compounded when abstractions are taken to be foundations of concrete reality and used to build further abstractions, taking us further and further from experiential concrete actualities. Universal abstractions can then start to take on independent lives of their own, importing qualities that do not exist, and when supported by infrastructures that allow mobility, they tend to occur more profusely and in increasingly varying places.

We live, increasingly, in an age where different entities connect and integrate with other entities – traffic systems being no exception – not *across* time and space (which is not constant), but by way of particular, and very technical, modes of connections that, themselves, *create* time and space (such as the Internet of Things). This requires conduits and corridors constituted from common languages and logics (Latour, 1986) or infrastructures of mobilization, such as described by Latour (1986) as "a regular avenue through space" (p. 7). The attraction towards mathematics in the development of these *infrastructures of technicity* and conduits of common languages lay in its technical efficiency and its character of neutrality (Whitehead, 1948). Whitehead (1948) states, "the point of mathematics is that in it we have always got rid of the particular instance, and even of any particular sorts of entities" (p. 22). This character of neutrality and universalism – the fact that no mathematical truths apply, for example, only to fish – means that it lends itself to universal formulaic constructions of abstractions that may then come to displace the particular.

The fundamental paradigms that form the foundations of many representational models for traffic research tools and technologies utilize the paradigm of *simple location*, such as global positioning systems (GPS), geographic information systems (GIS), and various traffic monitoring systems like intelligent transportation systems (ITS). These technologies and the paradigms that underpin them allow fundamental assumptions to be made about the nature of reality, often in contrast to how complex meaning emerges through our actions and experiences in the world.

Raper and Livingstone (1995) write that spatial modelling in the environmental sciences is often compromised due to the representational frameworks that the environmental models are embedded in. In their words, "The representational basis of the GIS is often allowed to drive the form and nature of the environmental model" (p. 359). Instead, they recommend that spatio-temporal referencing should be derived from the particular phenomenon, such as an object-oriented spatial modelling system. Massey (1999) considers that space should be seen as open and dynamic (p. 264), constantly in the process of being made. She notes that the following notions around space-time are increasingly being incorporated into human geography studies:

1   Space-time is relative and is defined in terms of the entities;
2   Space-time is relational, constituted through the operation of social relations; and
3   Space-time is integral to the constitution of the entities (the entities may be thought of as local time-spaces).

One needs look no further than the use of statistics as depictions of the current state of a complex system to understand the degree to which mathematics constitutes dominant frameworks of representation. For example, Vietnamese traffic researchers commonly make use of statistics, often pages of them, in order to depict the current state of traffic realities. These documents are sometimes even accompanied by a caveat noting the "statistical inconsistency" (Khuat & Le, 2011, p. 88) of the data, due to under-reporting of incidents and inherent structural problems such as endemic corruption related to data collection as well as issues around interpretation or communication of the data. As Nguyen, Chu, Nguyen and Thach (2013) note:

> These data are legally admitted as official. But unfortunately, the data source has five great shortcomings: lack of systematization, underreporting, incompleteness, incorrectness and inaccessibility. So, these data reflect the traffic safety situation incorrectly. (para. 2)

How did we get to a point where a scientific paper, attempting to communicate a realistic statement about nature, begins with the premise that the initial data, from which the interpretation then proceeds, are, mostly likely incorrect, and yet continues to follow through with the whole exercise? Furthermore, why would it be necessary to publish such an exercise that begins from a flawed premise other than to demonstrate familiarity with a scientifically acceptable method of analysis? These five 'shortcomings' that were treated as mere "noise" or roadblocks on the way to pure objectivity offer opportunities into better understanding the relationality and dynamics that give rise to the phenomenon. Instead of getting a phenomenology of traffic, the dominance of one particular representational model, somewhat divorced from concrete reality, forces us to settle for a phenomenology of matter-of-factness and an inaccurate and partial rendering of reality.

It is statistics, however, that remain the most readily available representation of the dynamics of the traffic in Ho Chi Minh City (HCMC). For example, from 2001 to 2014, the population of Vietnam increased from approximately 81 million to about 92 million, and during the same time period, the number of motor vehicle registrations increased from about 9 million in 2001 (Hsu, Sadullah & Dao, 2003) to 43 million in 2014 (number arrived at by the Vietnam Association of Motorcycle Manufacturers and the National Traffic Safety Committee, as cited in Le, 2015). Therefore, over this 13-year period, the situation went from only one in nine people owning a motorcycle to about one in two, a number that must also take into account children. Many of those who now owned motorcycles then moved to live in HCMC and the city almost doubled in population from 2001 to 2014 from 5.4 million to more than 8 million (Statistical Department of Ho Chi Minh City). In mid-2017, the mayor of HCMC, Nguyen Thanh Phong, suggested that there were 13 million people living in HCMC, a figure that included unregistered inhabitants from the provinces. He also cited official transportation figures that put the number of registered motorcycles in the city at 7.6 million (not including the 1 million or more bikes brought in by migrants), 700,000 cars, and 1,000 new vehicles entering the city's streets every day (as cited in Vu, 2017).

In 2017, it was estimated that 180 new cars were added to the streets of HCMC every day (Vinkenborg, 2017), and in 2018, the Vietnamese government eliminated all tariffs on cars (Completely Built Units) imported from ASEAN countries, a development that will most likely drastically increase car numbers on the roads in Vietnam. Passenger cars are still, very much, a statistical minority in HCMC, comprising, at the most, only about 8% of the traffic mix. While 180 cars a day seems like a lot, in a city of 10 million people, at 8%, it also seems like a little. However, the devil is not in the numbers, but in the details and the particulars, and the real impact of these cars on HCMC streets is better represented in more *matter-of-concern* ways, in how they are contingently, complexly relationally constituted on the streets of HCMC. Reality is complex, particular, and contingent, and so numbers might appear as objectivity, but such mathematical portrayals construct a barren landscape that exists in another realm from the subjective, emotional affectivity of what it actually feels like to live and work in such conditions.

Traffic research, such as computer models and simulations, and mathematical depictions of traffic, such as through statistics, are often embedded in relationalities, processes, and modes of thought that attempt to take concrete actual performances embedded in relationalities of experiential complexity and reconstruct them as matter-of-fact objects, embedded in paradigms and infrastructures of technicity, of mathematical algorithms, and mathematical physics, often with the outcome of producing an object divorced from the particularities of the original performance. Therefore, the fundamental question remains: are these kinds of research paradigms, so dominant in urban traffic research, appropriate and productive for the kinds of phenomena under analysis, and what other ways might be more suited for representing phenomena so value-laden and complexly, subjectively constituted?

References

Dreyfus, H. L. (2007). Philosophy 185 Heidegger [Audio podcast]. Retrieved from https://archive.org/details/Philosophy_185_Fall_2007_UC_Berkeley

Heidegger, M. (1977). *The question concerning technology, and other essays*. New York, NY: Harper & Row.

Heidegger, M. (2001). *Poetry, language, thought* (A. Hofstadter, Trans.). New York, NY: Harper Perennial Classics.

Hosinski, T. E. (1993). *Stubborn fact and creative advance: An introduction to the metaphysics of Alfred North Whitehead*. Lanham, MD: Rowman & Littlefield Publishers.

Hsu, T. P., Sadullah, A. F. M., & Dao, N. X. (2003). A comparative study on motorcycle traffic development of Taiwan, Malaysia, and Vietnam. *Journal of the Eastern Asia Society for Transportation Studies, 5*, 179–193.

Khuat, V. H., & Le, T. H. (2011). Education influence in traffic safety: A case study in Vietnam. *IATSS Research, 34*(2), 87–93.

Latour, B. (1986). Visualization and cognition. In E. Long & H. Kuklick (Eds.), *Knowledge and society: Studies in the sociology of culture past and present* (Vol. 6, pp. 1–40). Retrieved from http://www.bruno-latour.fr/sites/default/files/21-DRAWING-THINGS-TOGETHER-GB.pdf

Latour, B. (1999). *Pandora's hope: Essays on the reality of science studies*. Cambridge, MA: Harvard University Press.

Latour, B. (2004). Why has critique run out of steam? From matters of fact to matters of concern. *Critical Inquiry, 30*(2), 225–248.

Latour, B. (2005). *Reassembling the social: An introduction to actor-network-theory*. Oxford, United Kingdom: Oxford University Press.

Latour, B. (2008). What is the style of matters of concern. In *Two lectures in empirical philosophy*. Department of Philosophy of the University of Amsterdam, Amsterdam: Van Gorcum. Retrieved from http://www.bruno-latour.fr/sites/default/files/97-SPINOZA-GB.pdf

Latour, B. (2011). Some experiments in art and politics. *e-flux, 23*, 1–7. Retrieved from http://worker01.e-flux.com/pdf/article_217.pdf

Latour, B. (2014). How better to register the agency of things: Tanner lectures, Yale. Retrieved from http://www.bruno-latour.fr/sites/default/files/137-YALE-TANNER.pdf

Latour, B., & Woolgar, S. (1986). *Laboratory life: The construction of scientific facts*. Princeton, NJ: Princeton University Press.

Law, J., & Mol, A. (2001). Situating technoscience: An inquiry into spatialities. *Environment and Planning D: Society and Space, 19*(5), 609–621.

Le, H. (2015, February 6). Number of registered bikes exceeds 2020 vision. *The Saigon Times*. Retrieved from http://english.thesaigontimes.vn/39463/Number-of-registered-bikes-exceeds-2020-vision.html

Massey, D. (1999). Space-time, "science" and the relationship between physical geography and human geography. *Transactions of the Institute of British Geographers, 24*(3), 261–276.

McIntyre, L. (2010, April 15). *Mcintyre on Macintyre: A tribal dispute on the messiness of the social sciences* [PDF file]. Retrieved from https://www.bu.edu/cphs/files/2013/08/Comments_on_MacIntyre.pdf

Nguyen, H.D., Chu, M.H., Nguyen, T.L., & Thach, D.A. (2013). Vietnam Traffic Safety through Data of Health Sector: Further Study. *Proceedings of the Eastern Asia Society for Transportation Studies*. https://www.academia.edu/5794271/Vietnam_Traffic_Safety_through_Data_of_Health_Sector_Further_Study

Raper, J., & Livingstone, D. (1995). Development of a geomorphological spatial model using object-oriented design. *International Journal of Geographical Information Systems, 9*(4), 359–383.

Simondon, G. (2011). On the mode of existence of technical objects. *Deleuze Studies, 5*(3), 407–424.

Sloterdijk, P. (2012). Nearness and Da-sein: The spatiality of being and time. *Theory, Culture & Society, 29*(4–5), 36–42.

Statistical Department Ho Chi Minh City. (n.d.). Giải thích thuật ngữ, nội dung một số chỉ tiêu thống kê dân số và lao động [Explanation of terminology, of some statistical indicators on population and labour] [Fact sheet]. Retrieved from http://www.pso.hochiminhcity.gov.vn/c/document_library/get_file?uuid=bb171c42-6326-4523-9336-01677b457b13&groupId=18

Stengers, I. (2011). *Thinking with Whitehead: A free and wild creation of concepts*. Cambridge, MA: Harvard University Press.

Thrift, N. (2008). *Non-representational theory: Space, politics, affect*. New York, NY: Routledge.

Vinkenborg, M. (2017, January 6). Stuck in traffic: Opportunities for urban infrastructure development in Ho Chi Minh City. *Vietnam briefing*. Retrieved from https://www.vietnam-briefing.com/news/stuck-traffic-opportunities-urban-infrastructure-development-ho-chi-minh-city.html

Vu, V. (2017, August 17). Guess how many people are jamming into Saigon? Hint: It's as bad as Tokyo. *VNExpress*. Retrieved from https://e.vnexpress.net/news/news/guess-how-many-people-are-jamming-into-saigon-hint-it-s-as-bad-as-tokyo-3628742.html

Weber, T. (2016). Metaphysics of the common world: Whitehead, Latour, and the modes of existence. *The Journal of Speculative Philosophy, 30*(4), 515–533.

Whitehead, A. N. (1920). *The concept of nature: Tarner lectures delivered in Trinity College*, November 1919. Cambridge, United Kingdom: Cambridge University Press.

Whitehead, A. N. (1948). *Science and the modern world*. New York, NY: The New American Library of World Literature.

Whitehead, A. N. (1968). *Modes of thought*. London, United Kingdom: Collier-Macmillan.

Whitehead, A. N. (1978). *Process and reality: An essay in cosmology (corrected edition)*. New York, NY: The Free Press. Originally published in 1929 by Macmillan.

5 Complexity and computer modelling of traffic

The notion of 'complexity,' of what exactly it is, and of what use it may be is, itself, something of a complex dilemma (Corning, 1998; Érdi, 2008). In terms of being a theory, complexity is a *mathematical* theory, rather than a *scientific* theory, which seeks to find/create relational patterns that describe a system's complex behavior, emerging as visual forms (such as fractal patterns), using the tools of mathematics and the power of computers (Capra & Luisi, 2014). Capra and Luisi (2014) describe the focus on nonlinear dynamics in complexity science as representing:

> A qualitative rather than a quantitative approach to complexity and thus embodies the shift of perspective that is characteristic of systems thinking – from objects to relationships, from measuring to mapping, from quantity to quality. (p. 99)

In other words, complexity theory is a relational approach that attempts to reveal topologies and topographies that might otherwise go undetected by moving from quantitative data to 'qualitative' representations. However, some scholars, such as Hayles (1999), have noted that there remains the question of just what this qualitative image represents and the validity of the relationship between the output and the actual experiential world. This is mostly due to the fact that the process of moving from quantitative to qualitative modes often begins with high-level abstractions that disregard more intuitive modes of experience and then towards the application of further abstractions as explanations of concrete realities. Like the vague, amorphous entity known as 'culture,' 'complexity,' as a thing in the world (or in the mind, or somewhere in-between), is best considered through the particular and the contingent in actual concrete events, rather than through abstractions not based in the experiential world, which may take us further into the depths of abstraction and further away from the "'booming, buzzing confusion' of the real world" (Corning, 1998, p. 198).

For example, freeway and urban traffic is considered to be a *complex* phenomenon (Kerner, Klenov, Hiller, & Rehborn, 2006; Urry, 2004), displaying characteristics of complexity, and so many research approaches to traffic, such as computer modelling and simulations, often have a primary goal of engaging with and learning more about complexity itself. In the case of the present book, the *object* under analysis is traffic, but the *subject* of the investigation is really more about phenomenology and the nature of experience. Nevertheless, the present investigation remains committed to the task of unpacking the abstraction 'traffic' through its particular concrete experiences – working from the concrete, back to the abstract. In computer modelling, traffic can begin to look less like an experiential phenomenon and more like a mathematical object, making the object of the study, mathematics, rather than traffic. It is, perhaps, for this reason, that Schönhof and Helbing (2009) conclude, "the on-going controversial debate about the right modelling approach, however, shows that traffic science has still not reached a generally accepted theoretical framework" (p. 784).

When a thoroughly emotional, subjective, and experiential phenomenon, such as the urban traffic in Ho Chi Minh City (HCMC), is translated into neutral and objective languages and frameworks of representation for the sake of efficiency, amongst other goals, essential dimensions of the phenomenon – that make it what it is – are lost along the way. The geographer Olsson (1974) says, "our statements often reveal more about the language we are talking *in* than about the things we are talking *about*" (p. 53) and "the price for this admission ticket into mathematics is extremely high" (p. 53).

Complex systems are said to be complex because they display certain universal characteristics of complexity. According to Cilliers (2002), a celebrated complexity theorist, these include their having a large number of elements that interact with each other in dynamic, nonlinear ways, so that the system changes over time, and small causes or inputs can have large outcomes or results. Also, there exist feedback loops, whereby activities by entities can feedback into the system in positive or negative ways, stimulating or inhibiting further activities. Complex systems also produce emergent outcomes that are different in kind from their individual components, and each component or entity in the system responds only to local information and remains ignorant of the state of the system as a whole. Cilliers (2002) points out that for an entity to know the behavior of the whole system, all of the complexity of the system would need to be present in that single element. This is an interesting point, and it might be said that in Whitehead's scheme, entities, before they are actualized through limitations, before they become settled facts, do contain (or, at least, have access to) all of the complexity of the system, though the nature of this complexity is of fluid potential and therefore lacks structure. Complex systems are extremely sensitive to their histories, and their past is "co-responsible for their present behavior" (Cilliers, 2002, p. 4), so that any analysis of a complex system that ignores the dimension of time would be incomplete. Finally, these kinds of phenomena are open systems with borders that must be seen more as a matter of framing on the part of the observer, rather than innate features of the systems themselves.

Herman, Lam and Prigogine (1973) describe these complexities in terms of traffic behavior as:

> A game in which the rules are not fixed once for all, but vary according to circumstances. The modification of the rules implies a modification of behavior of the entire collectivity of participants. (p. 918)

In other words, what I have for breakfast before riding out into the traffic and where I bought it from might have nonlinear, knock-on effects that change the 'rules' or, in other words, the environment, in ways that no one could foresee. This implies a deep co-constitutional relationship between the entities and the environment. Urry (2005) likens complex systems to a maze in which the walls move and adjust to one's footsteps, and therefore, the footsteps need to also adjust to the wall movements. In other words, all of the actions of individual components, acting on immediate information around them (not necessarily physically proximate), contribute in defining the structure of the system, and the structure and its emergent outcomes also feedback into the system, then modifying the actions of the components at all "levels" of the system. Like a shape-shifting nest of Russian dolls, each layer is able to pop in and out of different temporalities and spatialities, and such nonlinear interactions make complex systems difficult to both control and to predict. For this reason, as Latour (2008) says, we are still "utterly unable to draw together, to simulate, to materialize, to approximate, to fully model to scale, what a thing in all of its complexity, is" (p. 12).

Complex systems are said to be different in nature to merely complicated systems. Whilst both kinds of systems might consist of large numbers of elements or components, if one had the right operating manual, a 'complicated' system could be taken apart and put back together again because the basic structure, principles, and relations would remain linear, that is, the same backwards and forwards in time. In fact, knowing all the pieces and where they are in time and space also makes it possible to predict the future state of the system (as well as the past), an idea known as Laplace's Demon, which is a thought experiment of an idealization of classical mechanistic science (Collier, 2011). Laplace's Demon, in attempting to demonstrate the principles of universal determinism, the foundational principles of which include reducibility, disjunction (separating the phenomenon into parts), and the truth criterion of simple universal laws, necessarily avoids notions such as emergence and holism and the dynamics of nonlinearity, which are characteristics of complexity (Morin, 2007).

Though complex systems are said to display certain 'universal' characteristics, such as autopoiesis (self-organization), emergence, nonlinearity, feedback loops, etc., even these universalities (this insistence on universal laws, itself, a hangover from classical mechanistic science?) require explanation through the particular and the concrete before gaining any real meaning. Cilliers (2002) says that complex systems defy formulations of universal, overarching theories, and that engagement with a complex system neces-sarily means engagement with "*specific* complex systems [emphasis added]" (p. ix). Therefore, the things that make a particular system 'complex' are unique to that system, so that "any system, whatever it might be, is complex by its own nature" (Morin, 2007, p. 10). What, then, makes a thing *complex by its own nature*? Nonlinearity and feedback loops refer to certain universal kinds of modes of relations, and yet, the ways in which these modes play out in the particular, the concrete, and the specific might involve a range of other kinds of modes of relations.

This means that the laws governing different complex systems are not eas-ily transferable across different kinds of systems, for example, from physical systems to sociotechnical systems. I might be able to drive through the city of my hometown – almost with my eyes closed – but how much of that knowl-edge is transferable to renting a car and driving as a newcomer in HCMC? Complexity lacks a universally accepted definition because, as Wells (2013) points out, "complexity is relevant relative to systems, interactions, observ-ers, and particular inquiries" (p. 32), suggesting that complexity itself is not an inherent property of a system, but is, instead, only "in the eye of the beholder" (comment made by Dan Stein, quoted in Corning, 1998). If the nature of complexity is different from phenomena to phenomena, from en-quiry to enquiry, from system to system, it might be questioned just how useful a universal definition of complexity is at all. Things seem 'complex' due to the manner in which a phenomenon is approached or the framing of a particular problem, what Corning (1998) calls "subjective complexity" or its meaning to a human observer (p. 198). Therefore, the observer is necessarily implicated and can never be taken out of the equation, the formulation, or the representation of a system, and for this reason, complexity may well be an inadequacy of language or a model, rather than an inherent property itself.

The notion of "wicked problems," a phrase first coined in a 1972 paper by Kunz and Rittel, is, more or less, another term for complexity. The term 'wicked' is used to refer to problems that seem to defy any definitive formu-lation and that resist clear-cut 'right' or 'wrong' categorizations, due to a contrasting range of perspectives and values. Framing wicked problems is problematic because the information required to understand the problem depends upon the questions being posed and the process used for solving it in the first place (Rittel & Webber, 1973). Wicked problems are 'wicked,' because they are embedded in complex systems and are multi-dimensional and ontologically layered, with many interdependencies that may result in initiatives and interventions leading to unforeseen and unpredictable conse-quences (Briggs, 2007).

For example, just the other week, I was riding across the Saigon bridge, not far from where I live, and I observed the wreckage of a motorcycle accident that must have occurred only a minute or two earlier. What was noteworthy amongst the wreckage was a motorcycle helmet that had been shattered into many different pieces across the road. Not so many years ago, a new law was introduced for the mandatory wearing of helmets for motorcyclists. Many helmets are produced without adherence to acceptable safety standards, but also meet the financial requirements of a large majority of traffic users in Vietnam. Wearing a helmet, even one that shatters on impact – possibly leaving shards of plastic in ones' head – might still give a motorcyclist some kind of psychological feeling of safety and lead to faster speeds or more reckless behaviour, an outcome that might never have been foreseen when the law was first introduced. Wicked problems are complex then, because their multi-dimensionalities are not easily corralled in practice, instead, flowing between the realms of the natural, environmental, cultural, social, political, economic, and technological dimensions of life. Depending on how the problem is framed, and depending on the perspective, there is always more than one single truth, question, or answer.

Also, things are complex due to the existence of 'unknown unknowns.' Lamprecht (1971) recounts a story of a skilled hunter, whose bullet unexpectedly misses the deer in his sights. On examination, the hunter sees a broken twig and surmises that a gust of wind moved the twig into the path of the bullet, thereby changing its trajectory. In other words, a different line of causality that first did not exist in the parameters of the situation intruded to rearrange the game. Deers and hunters and twigs and bullets and small gusts of wind are actual concrete things that are causally related and are also particular and contingent factors, rather than universal laws. The notion of 'unknown unknowns' is particularly relevant for traffic models, whose designers need to decide what to include in the 'game' and what to leave out, an impossible dilemma, given that seemingly insignificant and unrelated things can cause major outcomes over time.

Computer modelling, which has its roots in the science of complexity, complex adaptive systems theory, chaos theory, and thermodynamic dissipative systems theory, evolved, to a large extent, dependent on developments in computational power and computer technologies. There are two significant challenges in studies of complex systems, both of which are interrelated with each other and relevant to our discussion of computer modelling and simulations of traffic systems. These challenges are, first, the enormous number of possible variables involved in complex systems and, second, the co-constitutive nature of the relationship between entities and their environments. Cilliers, in the essay "Why We Cannot Know Complex Things Completely" (2016), says that most models of complex systems show general complex behaviour but not the particularities and contingencies of specific complex systems, which are important, causally speaking. In Lamprecht's example of the hunter, the story is told from the perspective (or percipient event) of the hunter. Should, however, the story be told from the point-of-view of the gust of wind, as an actant, the whole perspective of the event shifts in ways that require a complete reshuffle of the 'walls of that maze' that Urry spoke of.

Any model of the traffic in HCMC must include the social, cultural, technical, natural, political, economic, etc. dimensions of life, as even the mundane act of driving a car is simultaneously involved in more than one layer of the system, a system that remains implicated in the history of mathematical physics, engineering, and earlier versions of traffic systems, the oil and gas industry, as well as systems composed of signifiers of social status, the fashion industry, funeral practices, car-as-money-making, etc.

It is for the above reasons that the *kinds* of relations between components of the system are usually more important than the components themselves and that the boundaries of the systems – which are open systems exchanging energy with their environment – cannot be spatially conceived (Cilliers, 2001). Of course, for a system to be a system at all, there must be some conception of a boundary, but, as already noted, this is where the observer or framing of the enquiry becomes implicated. Having said that, as Cilliers points out, due to the nature of nonlinearity, each component is, in a virtual kind of way, as close to the 'boundary' as any other component (2001).

In January 2019, a speeding cargo truck ploughed through a large group of motorcycles waiting at a traffic light intersection on the outskirts of HCMC, killing four people and injuring many more, breaking motorcycles in half, and dragging some more than 150 metres (Hoang Nam & Huy Phong, 2019). The online video, which is as surreal as it is horrific, shows the truck, without warning and without slowing down, speeding through a tightly bunched group of about 30 motorcyclists waiting at the traffic lights. Moments after the truck had swept through the motorcycles, dazed and confused traffic users can be seen wandering amid the wreckage (what would time have felt like for them?). Watching the video, it is as though a kind of reality-snowplough had come through: one moment, reality was there, stable, dependable, made of enduring objects that make sense to us, and the next moment, it was gone, shifted, and erased.

Speaking with residents in HCMC about this event, a common question arose: what was the driver *thinking*? It seems to me that the driver was not so much thinking, but simply engaged in a kind of absorbed coping, an affective attunement that was profoundly influenced by a number of complex relations. The truck driver – who fled the scene on foot, as though the whole thing might just go away if he chose not to think about it too much – tested positive for heroin and large quantities of alcohol in his bloodstream (Saigoneer, 2019). This event is multidimensional in that it involves issues around social drinking (he had come from a party), unrealistic timelines for drivers (hence the stimulants). However, for me, the most significant factor in this event is that the driver imagined that he might *fit through a physical space that was simply not there*. In the video, there appears no intent by the driver to slow down, and the truck's transition from the dedicated truck lane into the motorcycle lane is smooth, as though the group of motorcycles was something that could just be passed through or *jostled* through. The driver (who, we should remember has spent much of his life riding a motorcycle) seemed to imagine a pathway where no pathway existed, as though he were contained in a much smaller body, one that could somehow fit through the mere centimetres of space that existed between the motorcycles that were bunched together at the traffic lights, a mental representation of his hybrid self that was perhaps also heightened by the drugs he had taken.

What is also significant about this event is that though it is in the past, forms endure from the event, continuing to influence and creating structural changes to the system as a result. One HCMC motorcyclist confided to me, after watching this video online, she now obsessively looks in her rear-view side mirror when waiting at traffic lights, in fear that a truck might plough through from the back.

There are also institutional arrangements regarding time that impact the behaviour of truck drivers. Trucks are not allowed to drive in the city limits between 6 am to 8 am and 4 pm to 8 pm, and Phúc, who is a professional driver of a small delivery truck, told me that he drives with the constant fear of being caught in a traffic jam, getting stuck somewhere, and then having to wait until after 8 pm, so that he can drive home and spend time with his family. He said, "Trucks normally follow the rules but when the peak time is coming, they also want to go home with their family so they have to drive fast not to get stuck within the peak time."

Just recently, I was a passenger in a Grab taxi car, driving along the main Hanoi highway that also leads to the port, when we found ourselves sandwiched in a sea of enormous container trucks. My driver, who was a seasoned professional and has driven for several international companies, said to me, "Look at their faces," gesturing towards the truck drivers all around us. They were all young. My driver said the average age of these drivers – who, we should remember, are responsible for driving a potentially devastating piece of mobility – is between 25 and 30 years of age. Then he hit me with another fact: the police routinely test truck drivers for drugs and eight out of ten truck drivers test positive, commonly for heroin or 'ice.' The driver also told me that it is very common for accidents to occur in which motorcycles go under these large trucks.

The following week, I was again on the Hanoi Highway, and was able to view the highway from another perspective, from my motorbike. The motorcycle lane is separated from the truck lanes by concrete barriers, but that doesn't stop them from travelling in the motorcycle lane at times. The trucks seemingly shrieked from the treatment they were getting, as overly competitive drivers repeatedly jumped on brakes and accelerators, constantly changing lanes, often straddling lanes, and leaving the barest of space between their vehicles and the truck in front; these trucks are enormous and yet they are driven with the sudden movements more akin to the affordances of motorcycles.

The air was filled with truck exhaust fumes and the sounds of horn blasts, wielded like some audio battering rams. The scene reminded me of a stampede of wild animals running away from some kind of danger, as though in panic. At one point along the road, there is a traffic light intersection where the entire fleet of trucks passes diagonally across the motorcycle lane in order to take an overpass to the port. During only my first time at this intersection, I witnessed three large container trucks run the red light, one of them just managing, with a scream of brakes, to stop before ploughing through two motorcycles that became stranded in the middle of the intersection, as the motorcycles' paths were blocked by the continuous wall of trucks that were also stuck across the intersection. There are many instances in HCMC whereby larger vehicles, such as buses, trucks, and cars, are required to pass through streams of motorcycles travelling in the motorbike lane, a situation that is invariably dangerous, perhaps more so for the motorcyclists.

In terms of the experience of mobilities, the Hanoi Highway, from District 2 to District 9, is, euphemistically speaking, fairly unpleasant (a judgement that, of course, depends on one's criteria of a 'pleasant' mobilities experience, and I see many motorcyclists who seem completely at ease in this environment). Nevertheless, a brown haze, perhaps from the exhaust fumes of so many vehicles, sits over this part of the city, which catches in my throat, despite the facemask I am wearing. It is extremely noisy, dangerous, and unpredictable, with vehicles of all kinds scrambling and competing against each other.

How can computer models hope to accommodate such complexities, material heterogeneities, and the varying different kinds of relationships that constitute a change in the very *nature* of phenomena from place to place? Are models, in fact, trying to replicate reality in the first place, and, if not, what might be their aims? Computer modelling generally falls into two different paradigms: the *rule-based symbol system* and the *connectionist* model (Cilliers, 2002). Rule-based systems use tokens or symbols that can be combined into patterns through sets of rules that allow or disallow configurations or actions. The symbols can be then made to represent particular entities, and the interpretation of the symbols depends on the rules governing the system (Cilliers, 2002). The problem with the rule-based approach is that the structure is pre-determined, resulting in two levels of the system: the symbols and the rules. Olsson (1974) says that Western paradigms are implicitly embedded with a dualistic logic where something must be seen as a 'this' or a 'that', a 'yes' or a 'no', and says that these paradigms manifest in material forms. In rule-based models, entities and their environments are clearly demarcated, and the rules must be spelled out in advance. On the other hand, connectionist models attempt to collapse this dichotomous reality by allowing the structure to develop more in a co-constitutive relationship with the entities, similar to how learning occurs in the brain through self-organization: relations get strengthened the more they are used and the structure develops in this way.

Computer simulations share many overlapping interests with artificial intelligence research, and Dreyfus suggests that we look towards Heidegger and Merleau-Ponty as guides towards developing better models for artificial intelligence. As Dreyfus (2007) says, "Our everyday coping [how people behave] couldn't be understood in terms of inferences from symbolic representations . . . [and] it can't be understood in terms of responses caused by fixed features of the environment" (p. 1142). Dreyfus suggests that we should make models more Heideggerian and find ways to engage with the complex relational constructions of meaning, viewing space as more existential, emerging through practice, therefore being directed and directional (Heidegger, 1996). Dreyfus (2007) argues that the fundamental obstacles standing in the way of developing successful models of complex systems are the incorporation of relevance and significance. When something changes in the world, how might an entity 'decide' which facts are *relevant* and which rules or sets of rules to attend to in any given situation? In a model containing only facts, controlled by simple rules of interaction, how are we to incorporate dimensions of relevance and significance, which are fundamental to any kind of existential being? At the primordial level of unconscious coping and attunement, things are both a 'this' and a 'that.' Propositions are a hybrid mix of possibilities and the limiting conditions of facts, amid a fusion of sensuous and non-sensuous perception (Hosinski, 1993).

Paradigms of reductionism and disjunction and the inability for things to be simultaneously both a 'this' and a 'that' are problematic for computer modelling of complex systems. Hayles (1999), in her book *How We Became Posthuman*, says, "when we make moves that erase the world's multiplicity, we risk losing sight of the variegated leaves, fractal branchings, and particular bark textures that make up the forest" (p. 12). Chris Urmson, who was the head engineer and who helped build the code for Google's autonomous vehicle, said (in a rather reductionist manner), "it's really just two numbers at the end of the day. . . so how hard can it really be?" (Urmson, 2015). He was referring to the process by which complex traffic environments are converted into algorithms in order for the Google autonomous car to make sense of its environment and to navigate within it. The "two numbers" he refers to are 'either/or' decisions, turn left or turn right, accelerate or brake. In these algorithmic representations, traffic objects show up on the visual display as geometric forms: purple boxes being other vehicles, red boxes being cyclists, etc. Chris went on to say that the computer has to predict what is going to happen from this data and, to do that, he said, "we really need to know what everybody is thinking." The complexity involved in any urban traffic situation is obviously enormous and should we ever develop a driverless car that knows what everyone is thinking and, from that information, is able to predict the future state of the system, it would be not so much a car, but a God.

In Vietnam, things can often seem to exist, simultaneously in practice, as both a 'this' and a 'that' (for example, the country's economic model is a practical fusion of socialism and capitalism). An online news site (Tuoi Tre News, 2016) recently featured an opinion piece written by an Indian expat living in HCMC that described – to his amazement – the existence

of particular traffic light intersections located around HCMC that operate using programs that display green traffic lights for streams of traffic
coming from multiple different directions simultaneously. According to the
writer, this results in vehicles from all directions "scrambling" (para. 1) to
get through the intersection at once and often getting stuck in the middle.
It must be said that traffic users in HCMC are very familiar with this kind
of fluidity and flexibility, but this brand of chaos can still be surprising for
users who are familiar with traffic in Indian cities.

How might a computer model contend with this level of heterogeneous
complexity? Resnick, a computer scientist at MIT, and author of the book,
Turtles, Termites and Traffic Jams (1997), says that the goals of his computer modelling experiments are more about stimulating thought, what he
calls *explorations in microworlds*, rather than attempting to actually simulate real-world processes or systems because, whilst they are *inspired* by
real-world phenomena, they are vastly simplified. Byrne (2011), a sociologist
and complexity theorist, says that computer models are "worse than use-
less . . . actually negative in their impact, if they are asserted as some proper
'scientific' account of complex social reality" (p. 154), and Helbing (2012),
an expert in the modelling of complex systems, says, "the main purpose of
models is to guide people's thoughts" (p. 7).

There is nothing inherently wrong with computer models, of course, but a
problem does arise when the sets of abstractions they are embedded in, and
the further abstractions that emerge from them, are then used to explain
and displace the real concrete experiential world, in the manner of Whitehead's fallacy of misplaced concreteness. As Hayles (1999) points out, all
good theorizing should move from "the world's noisy multiplicity" (p. 12)
towards more simplified abstractions, but when we move from simplicity
towards multiplicities, using the abstract to explain the concrete and privi-
leging the abstract over concrete instantiation – a move Hayles refers to as
the "Platonic forehand" (p. 12) – this results in the bifurcation of nature.
Abstractions are reliable footholds in complex reality and are necessary for
meaning. They allow us to quickly skip over rivers of otherwise unfathom-
able complexity, like stepping-stones across swirling waters of flux and possibility. But 'misplacing' or displacing concreteness tends to ignore all those
rich complex forces, the eddies, and swirls, without which, these stepping-
stone abstractions would never achieve their unique shapes in the first place.
The attraction towards abstractions lay in their permanence and simplicity,
residing, as they do, outside of the flux and change of events, but, as Stengers
says, "what is abstract can never explain what is concrete" (2011, p. 99) because it is simply moving in the wrong direction.

Traffic is complex because it is constructed of technologies, patterns of
practices (Schatzki, 2010; Shove, Pantzar, & Watson, 2012), materiality and
mediated actions (Latour, 1999), and "collective forms of perception and
sensation" (Reckwitz, 2012, p. 242), by way of temporal–structural shapes
and artefact-space structuration (Reckwitz, 2012), as well as regulations and

policies (Merriman, 2006), and is also profoundly influenced by its unique circumstantial and historical origins (Bennett, 2005). Any description of traffic that attempts to truly engage with its unique complex and essential nature cannot exclude any of these dimensions. Whilst the dimensions listed above might be considered universal across all traffic systems, they only represent the bare bones of what constitutes traffic and do not account for the particular and "contingent modalities of connection" (Oppenheim, 2014, p. 393) that constitute the character of the system. As Whitehead (1978) says, "these subjective ways of feeling are not merely receptive of the data as alien facts; they clothe the dry bones with the flesh of a real being, emotional, purposive, appreciative" (p. 85).

Traffic research, when utilizing mathematical algorithms, is able to produce complex outcomes, but such outcomes are formed through processes and protocols, and constituted with relationalities that are often extremely divorced from the actual performances of contingent, place-based concrete traffic realities. The particular 'bark textures,' the modes, and relationalities of *experiential complexity* differ from the matter-of-fact reductionist relations and outcomes of processes of computer models that engage with complexity. As humans develop tacit skills and knowledge in traffic, attuned to embodied awarenesses, meaningful abstractions still arise, but the processes by which this occurs and the relations by which they endure are different from the protocol-bound processes that might occur in computer models. The abstractions that emerge from relations of experiential complexity retain the essential richness, subtleties, and multiplicities of affect and atmosphere, where something can be both a 'this' and a 'that,' a 'subject' and an 'object,' and an entity and environment. Hayles (1999) notes that this form of reductionist theorizing has been around a long time, but due to the technological developments of computation, is now in a new form. She writes, "this move starts from simplified abstractions and, using simulation techniques such as genetic algorithms, evolves a multiplicity sufficiently complex that it can be seen as *a world of its own* [emphasis added]" (p. 12).

The proliferation of these kinds of technologies means that these 'worlds' that Hayles refers to, borne from algorithms and processes that move from simplicity to mathematical complexity, are worlds within which we, in our existentialist being, also dwell. We are intimately embedded in these modes and relationalities, constructing their own kinds of infrastructures in the world, with their own particular essences, aesthetics, and style. As many scholars have commented, not least of all, Whitehead, Latour, Thrift, and Harman, this bifurcated view of reality has fundamental implications on the developments of different kinds of awarenesses as well as in the creation and maintenance of infrastructures imbued with particular kinds of logics and aesthetics (Harman, 2005; Whitehead, 1948). Thrift (2008) has commented that these developments result in new kinds of spatialities that are "both template and font" (p. 23) for our cognitive processes, changing structures, and patterns in ways equally as influential as the growth of writing.

One possible effect of the bifurcation of nature, according to Whitehead (1948), is described in the following:

> The disadvantage of exclusive attention to a group of abstractions, how-ever well-founded, is that, by the nature of the case, you have abstracted from the remainder of things . . . In so far as the excluded things are important in your experience, your modes of thought are not fitted to deal with them. (p. 58)

The bifurcated world of traffic research, with complex outcomes emerging from representational models embedded in mathematical algorithmic frameworks on one side, and the experiential complexity of the phenomenological experience of traffic on the other, creates an incompatible chasm between *thinking* about traffic and *doing* traffic, the danger being an ever-widening bifurcation and the possible importation of "a mere procedure of mind. . . transmuted into a fundamental character of nature" (Whitehead, 1920, p. 16).

Helbing (2012) writes:

> The dynamics of traffic flows can be mathematically well understood. Nevertheless, one cannot exactly forecast the moment in which free traffic flow breaks down and congestion sets in, and therefore, one cannot forecast travel times well. (p. 8)

The above quote highlights three fundamental aspects in traffic computer models that can be problematic: the aims (prediction towards efficiency), the frameworks of representation and the language in which they are embedded (mathematical algorithms), and processes that aim towards the objectification and universalization of knowledge. One of the aims of computer simulations remains the prediction of future states of complex systems, often with the further aim of then controlling some behaviour (Helbing, 2012), yet complexity theorists repeatedly point out that complex systems cannot be predicted or controlled (Byrne, 2011; Cilliers, 2002; Emmeche, Køppe, & Stjernfelt, 1997; Wells 2013). Helbing (2012) writes that the random effects of history-dependent dynamics make it less than an exact science because the high levels of nonlinear interactions mean that small factors can produce large outcomes. Yet, despite this, the application of the kinds of knowledge derived from such studies, to be used in intelligent transportation systems, for example, continues to be embedded in goals of better forecast and prediction of traffic systems towards better control and more efficient flow.

The frameworks of representation of computer models also dictate, to a large extent, other aspects of the research process, such as the tools and methods for the collection of empirical data to be then incorporated into the models. For example, from the outset, complex concrete realities may be reduced and mathematicized, such as when as traffic detectors are employed that measure numbers of vehicles, recording when a vehicle crosses the detector and measuring its speed (Kerner & Rehborn, 1996). Also, traffic models increasingly utilize entire matter-of-fact infrastructures such as GPSs and other subsystems of intelligent transportation systems (Kerner et al., 2006), all of which are embedded in paradigms of simple location. These data collection methods, themselves, then act as mediators and filters, translating and interpreting empirical data into certain styles and logics.

Metaphors that draw upon the dynamics of other natural physical forms such as fluid mechanics and gas kinetics are also commonly imported into traffic computer models and simulations. The dynamics of certain systems in the physical sciences such as gases and fluids are, like traffic systems, similarly described as complex systems. In the search for universal laws, the importation of notions borrowed from the physical sciences into traffic models is commonplace. However, according to Daganzo (1995), this can lead directly to two fundamental problems in traffic models: it "ignores the special nature of traffic particles, and it leads to strange predictions" (p. 280). Daganzo (1995) cites how the importation of such abstractions can lead to predictions of negative flows and negative speeds, for example, of all of the traffic going the wrong way, which we know, from our experience in traffic, does not occur. Daganzo (1995) suggests that traffic model predictions should be able to withstand what he calls a "mental test" (p. 281). In other words, we should reflect on our concrete experiences in traffic in order to consider the possibility of such a predicted scenario actually occurring in the concrete world.

There are still ongoing debates about the right kinds of modelling approaches and incompatible divides between empirical data and traffic models. The chasm that exists between the particularities of experiential traffic, on one side, and traffic models, on the other, as well as the problematic dichotomous ontology of environment and entity, suggests that traffic science frameworks are becoming increasingly embroiled in the bifurcation of nature. Herman (1991) has made comment that computer models have "far too little reliance on data, the language of nature, in formulating models for the systems and processes of the deepest importance to us human beings" (p. 310). Ideally, the 'language of nature' should come from the particular phenomenon itself, as *given*, in whatever mode that may be. In order to try to bridge the gap between empirical observation of traffic and models, researchers sometimes incorporate what Helbing describes as "non-mathematical and non-algorithmic" descriptions of reality in the form of narratives (Helbing, 2012, p. 4), and, as scholars have noted, narratives are important for revealing multiple levels of agency (Uprichard & Byrne, 2006).

Daganzo, when referring to the laws governing the behaviour of entities in a computer traffic model, said:

> These laws would have to prescribe that a slow car should be virtually unaffected by its interaction with faster cars passing it (or queuing behind it) . . . and that interactions do not change the "personality" (aggressive/timid) of any car. (1995, p. 279)

Daganzo (1995) makes the claim that "unlike molecules, vehicles have personalities (e.g., aggressive and timid) that remain unchanged by motion" (p. 279). Whilst Daganzo is absolutely right to introduce a subjective, value-oriented distinction between fluid particles and molecules and real humans driving cars, and that vehicles may have personalities, the idea that these personalities remain constant regardless of changes in motion or any other change in interaction does not correspond with common sense or common experience. For example, who has not felt the presence of an impatient person standing behind them in a queue, or the revving or particular timbre of a motorcycle or car engine, as affected by the temperament of the driver? Affect, which is such a fundamentally important dimension in traffic, and is the 'language of nature,' seems to be somewhat incompatible with the languages and tools of computer modelling and simulations.

Whilst some macroscopic computer traffic models are "phenomenologically motivated" (Schönhof & Helbing, 2009, p. 784), the language and the models of representation they work within remain algorithmic, which is problematic for integrating the emotional and aesthetic dimensions of people's experiences in traffic into the models. The problem then arises as to what forms and languages may constitute an aesthetic of 'matters of concern' or 'language of nature.' Helbing and Molnár (1995) attempted to incorporate more value-laden, affective, and aesthetic dimensions into computer traffic models – what they referred to as "social forces" – by adapting Lewin's social field theory (p. 4282). The social forces in these models are based on the probability of a sensory stimulus causing a certain behavioural reaction based on emotions or affect; i.e., we are either attracted to or repelled by something in the traffic environment. However, in order to incorporate this kind of subjective, aesthetic, affective, value-oriented realm into models, experiential complexities were translated into the language of the model of representation used in the models. This means that these affective forces are forced into a kind of lowest common denominator and into a technical language that is different in nature and incompatible with the emotional, subjective dimension of experience, from which arises significance and relevance as well as aims and goals. This disjunction is illustrated in how a negative subjective experiential feeling is described in the following terms:

> We will assume that the repulsive potential $V\alpha\beta(b)$ is a monotonic decreasing function of b with equipotential lines having the form of an ellipse that is directed into the 4 direction of motion. (Helbing & Molnár, 1995, p. 4283)

To the non-specialist, this is, indeed, a strange articulation of an affective and emotionally felt experience, though such algorithmic constructions might be filled with meaning and even emotion for traffic model designers. However, the exercise seems to only highlight the chasm that exists between the abstract and the actual, and suggests the need for the development of new tools and methods that are built up from the contingent modes of relations, rather than from universal laws, in order to better attend to the aesthetics and character of things in traffic, in the languages of their given nature.

References

Bennett, J. (2005). The agency of assemblages and the North American blackout. *Public Culture, 17*(3), 445–465.

Briggs, L. (2007). Tackling wicked problems: A public policy perspective. Canberra, Australia: Australian Government, Commonwealth of Australia. Retrieved from https://www.apsc.gov.au/tackling-wicked-problems-public-policy-perspective

Byrne, D. S. (2011). *Applying social science: The role of social research in politics, policy and practice.* Bristol, United Kingdom: The Policy Press.

Capra, F., & Luisi, P. L. (2014). *The systems view of life: A unifying vision.* Cambridge, United Kingdom: Cambridge University Press.

Cilliers, P. (2001). Boundaries, hierarchies and networks in complex systems. *International Journal of Innovation Management, 5*(2), 135–147.

Cilliers, P. (2002). *Complexity and postmodernism: Understanding complex systems.* London, United Kingdom: Routledge.

Cilliers, P. (2016). Why we cannot know complex things completely. In R. Preiser (Ed.), *Critical complexity: Collected essays* (pp. 97–103). Berlin, Germany: De Gruyter.

Collier, J. (2011). Holism and emergence: Dynamical complexity defeats Laplace's demon. *South African Journal of Philosophy, 30*(2), 229–243.

Corning, P. A. (1998). Complexity is just a word!. *Technological Forecasting and Social Change, 59*(2), 197–200.

Daganzo, C. F. (1995). Requiem for second-order fluid approximations of traffic flow. *Transportation Research Part B: Methodological, 29*(4), 277–286.

Dreyfus, H. L. (2007). Why Heideggerian AI failed and how fixing it would require making it more Heideggerian. *Artificial Intelligence, 171*(18), 1137–1160.

Emmeche, C., Køppe, S., & Stjernfelt, F. (1997). Explaining emergence: Towards an ontology of levels. *Journal for General Philosophy of Science, 28*(1), 83–119.

Érdi, P. (2008). *Complexity explained.* Berlin, Germany: Springer.

Harman, G. (2005). *Guerrilla metaphysics: Phenomenology and the carpentry of things.* Chicago, IL: Open Court.

Hayles, K. (1999). *How we became post human: Virtual bodies in cybernetics, literature, and informatics.* Chicago, IL: University of Chicago Press.

Heidegger, M. (1996). *Being and time: A translation of sein und zeit* (J. Stambaugh, Trans.). New York: State University of New York Press.

Helbing, D. (Ed.). (2012). *Social self-organization: Agent-based simulations and experiments to study emergent social behavior.* Zurich, Switzerland: Springer.

Helbing, D., & Molnár, P. (1995). Social force model for pedestrian dynamics. *Physical Review E, 51*(5), 4282–4286.

Herman, R. (1991). Traffic dynamics through human interaction: Reflections on some complex problems. *Journal of Economic Behavior & Organization, 15*(2), 303–311.

Herman, R., Lam, T., & Prigogine, I. (1973). Multilane vehicular traffic and adaptive human behavior. *Science, 179*(4076), 918–920.

Hoang Nam. & Huy Phong. (2019, January 2). Four killed, 16 injured as truck crashes into motorbikes in Vietnam. *VNExpress*. Retrieved from https://e.vnexpress.net/news/news/four-killed-16-injured-as-truck-crashes-into-motorbikes-in-vietnam-3862707.html

Hosinski, T. E. (1993). *Stubborn fact and creative advance: An introduction to the metaphysics of Alfred North Whitehead*. Lanham, MD: Rowman & Littlefield Publishers.

Kerner, B. S., Klenov, S. L., Hiller, A., & Rehborn, H. (2006). Microscopic features of moving traffic jams. *Physical Review E, 73*(4), 1–16.

Kerner, B. S., & Rehborn, H. (1996). Experimental features and characteristics of traffic jams. *Physical Review E, 53*(2), R1297–R1300.

Kunz, W., & Rittel, H. W. (1972). Information science: On the structure of its problems. *Information Storage and Retrieval, 8*(2), 95–98.

Lamprecht, S. P. (1971). Contingency in nature. *Philosophy and Phenomenological Research, 32*(1), 1–14.

Latour, B. (1999). *Pandora's hope: Essays on the reality of science studies*. Cambridge, MA: Harvard University Press.

Latour, B. (2008). A cautious Prometheus? A few steps toward a philosophy of design (with special attention to Peter Sloterdijk). In *Proceedings of the 2008 annual international conference of the design history society* (pp. 2–10). Retrieved from http://www.bruno-latour.fr/sites/default/files/112-DESIGN-CORNWALL-GB.pdf

Merriman, P. (2006). 'Mirror, signal, manoeuvre': Assembling and governing the motorway driver in late 1950s Britain. *The Sociological Review, 54*(1_suppl), 75–92.

Morin, E. (2007). Restricted complexity, general complexity. In C. Gershenson, D. Aerts, & B. Edmonds (Eds.), *Worldviews, science and us: philosophy and complexity* (pp. 5–29). Singapore: World Scientific.

Olsson, G. (1974). The dialectics of spatial analysis. *Antipode: A Radical Journal of Geography, 6*(3), 50–62.

Oppenheim, R. (2014). Thinking through place and late actor-network-theory spatialities. In P. Harvey, E. C. Casella, G. Evans, H. Knox, C. McLean, E. B. Silva, … K. Woodward (Eds.), *Objects and materials: A Routledge companion* (pp. 391–398). Oxford, United Kingdom: Routledge.

Reckwitz, A. (2012). Affective spaces: A praxeological outlook. *Rethinking History, 16*(2), 241–258.

Resnick, M. (1997). *Turtles, termites, and traffic jams: Explorations in massively parallel microworlds*. Cambridge, MA: MIT Press.

Rittel, H. W., & Webber, M. M. (1973). Dilemmas in a general theory of planning. *Policy Sciences, 4*(2), 155–169.

Saigoneer. (2019, January 3). Drunk driver plunges container truck into 25 motorbikes at traffic light, killing 4. *Saigoneer*. Retrieved from https://saigoneer.com/vietnam-news/15414-drunk-driver-plunges-container-truck-into-25-motorbikes-at-traffic-light,-killing-4

Schatzki, T. (2010). Materiality and social life. *Nature and Culture, 5*(2), 123–149.

Schönhof, M., & Helbing, D. (2009). Criticism of three-phase traffic theory. *Transportation Research Part B: Methodological, 43*(7), 784–797.

Shove, E., Pantzar, M., & Watson, M. (2012). *The dynamics of social practice: Everyday life and how it changes*. London, United Kingdom: Sage.

Stengers, I. (2011). *Thinking with Whitehead: A free and wild creation of concepts.* Cambridge, MA: Harvard University Press.

Thrift, N. (2008). *Non-representational theory: Space, politics, affect.* New York, NY: Routledge.

Tuoi Tre News. (2016, May 9). Make traffic lights smarter to relieve Ho Chi Minh City congestion: Expat. *Tuoi Tre News.* Retrieved from https://tuoitrenews.vn/city-diary/34697/reduce-congestion-in-ho-chi-minh-city-make-traffic-lights-smarter-say-readers

Uprichard, E., & Byrne, D. (2006). Representing complex places: A narrative approach. *Environment and Planning A, 38*(4), 665–676.

Urmson. C. (2015, June 26). *Chris Urmson: How a driverless car sees the road.* Lecture [Podcast]. Retrieved from https://www.youtube.com/watch?v=tiwVMrTLUWg

Urry, J. (2004). The 'system' of automobility. *Theory, Culture & Society, 21*(4–5), 25–39.

Urry, J. (2005). The complexity turn. *Theory, Culture & Society, 22*(5), 1–14.

Wells, J. (2013). *Complexity and sustainability.* London, United Kingdom: Routledge.

Whitehead, A. N. (1920). *The concept of nature: Tarner lectures delivered in Trinity College,* November 1919. Cambridge, United Kingdom: Cambridge University Press.

Whitehead, A. N. (1948). *Science and the modern world.* New York, NY: The New American Library of World Literature.

Whitehead, A. N. (1978). *Process and reality: An essay in cosmology (corrected edition).* New York, NY: The Free Press. Originally published in 1929 by Macmillan.

6 The character of character

The spirit of this book is such that traffic is viewed as a way into philosophy, while philosophy promises tools towards a deeper (or, at least, different) understanding of the nature of traffic and towards the development of models of representation more aligned with its modes of givenness. In attempting to understand the nature of experience in general, the deceptively simple, but variously entangled entities, we know as mundane 'objects,' feature as the chief protagonists. In exploring traffic and its objects, it is necessary to expand further the scope of the term 'object' and the possibilities of the *being* of objects beyond mere physicality and beyond their presentations as encountered by way of sensory perception; what Whitehead (1985) calls, the "show of our own bodily production" (p. 44). The reason being, that neither our practical activities with objects, nor our theoretical, objective analyses of them ever give us access to the 'real' thing; as Harman (2007) notes, "We distort when we see, and distort when we use" (p. 193). Instead, according to Harman, we are presented with a kind of caricature, as our 'seeings' and 'doings' with objects are, to a large degree, virtual. Latour (2008) refers to these distortions and the products of sensory perception as the *phantasmagoria* (p. 11), a realm of abstractions accessed through sense data perception and in our instrumental dealings with things.

In our journey towards ontologies that better capture the deeper complexities of the structures and stratifications that form our experiential realities, we have to start somewhere however. The distortions and caricatures of our objects suggest modes of *givenness* and present opportunities that provide access to the hidden phenomenological structures of experience, such as what Heidegger (1996) calls the realm of *tool-being* and what Massumi (2008) calls "lived relations" (p. 4), that lie beyond the curtain of immediate sensory perception, beyond the *phantasmagoria*, where phenomenology meets metaphysics. It is for this reason that Whitehead begins with *human* access, but then extends those structures of 'experience' beyond the limiting perspectives of an anthropologically centred universe. Brown (2001) says, "We look through objects [but we] only catch a glimpse of things" (p. 4). This glimpse signals that in our experiences with objects, we retain a vague awareness of the existence of unresolved possibilities, that something was missed, was not quite *captured* or closed off; an excess or openness that flows beyond the boundedness of our meaning-making. This excess, which exists as potentiality, also suggests a more virtual, 'out-there,' affective kind of existence, such as what Massumi (1995) calls "the autonomy of affect" (p. 83).

Whitehead's scheme – though he doesn't necessarily employ the term 'affect,' preferring, instead, his own specific vocabulary, such as 'prehension' and 'feeling' – presents an ontology, where the autonomy of affect is central to an understanding of the 'event,' to our experiences with multiplicities of actual entities, and to the processes through which aims, goals, and decisions arise. In addition, Whitehead (1978) also clearly states that his philosophy of organism is closely aligned with Spinoza's notion of substance, which is one of *affectiones substantiae* (p. 7). In doing so, Whitehead brings into view, a universe built up from fluid processes, simple, and complex events, and a notion of 'substance' and objects that turns the traditional views of such things on their head. In Whiteheadean terms, a mundane object such as a car, a bus, a truck – or myself, for that matter – is, at its most fundamental level, a series of events and therefore always in a state of *happening* (Shaviro, 2009). This suggests that objects, whose constitution we usually regard as relatively stable and constant, have a nature that is, in fact, more the opposite. Whitehead (1978) says, "an ordinary physical object, which has temporal endurance, is a society" (p. 35). Therefore, what we usually think of as a single independent entity, is, for Whitehead, a *society* or *nexus*, an event, or, more precisely, a series of events, made up of the most fundamental of entity/events termed *actual entities* or *actual occasions*, which he considers the building blocks of nature. The entity we know as 'car' is then viewed as a relational event/thing, a "multiplicity of definite actual entities" (Whitehead, 1978, p. 62), and, equally, a "plurality of processes" (Hooper, 1941, p. 285). A physical object is described as a collectivity of actual entities that coalesce in some way, contiguous in space and time (Shaviro, 2007a), and related by some commonality, by, as Whitehead (1978) says, "any such particular fact of togetherness" (p. 20). This suggests that the key to understanding the nature of objects lays in their formulation through their particular modes of connectedness or togetherness, a perspective shared with Heidegger's notion of the *gathering* and Latour's network theories.

These primordial, single, and simple 'event/vibrations' called *actual entities* form the 'bedrock' of reality, though a 'bedrock' of the molten consistency of on-going process. This fluid consistency comes about due to the fact that its fundamental *becomings*, actual entities, are so "swiftly transitory" (Hooper, 1941, p. 286), that they perish as soon as they come into existence, meaning that they are, in a sense, simultaneously both a process and a 'thing.' Hooper (1941) describes actual entities or occasions as like a "vibration" in the physical world (p. 286), and Whitehead (1978) refers to them as "drops of experience, complex and interdependent" (p. 18). Hooper (1941) describes the experience of an actual entity thus:

> In the realm of conscious life an actual occasion is the passing experience of a pleasure or pain, the experience of an emotion, such as anger or fear, or an aesthetic thrill of delight evoked by the contemplation of a beautiful object. (p. 286)

It is important to realize, however, that though we might feel these 'vibratory' affective entities or relations, we feel them complexly and vaguely in the form of what Whitehead (1985) calls *causal efficacy*, yet to be closed off or categorized into a specific 'this' or 'that' emotion or specific feeling. Emotion, like affect, is a complex feeling and encompasses multiplicities of actual entities and their relations, and it remains an unsettled question as to whether we experience any individual identity of a single actual entity, and what, exactly, the nature of such an experience might be. Garland (1982) says that there is much in Whitehead's writings that makes clear his intentions to state that we have direct experience of other actual entities as individuals. In fact, Chappell (1961) contends that for Whitehead to rationally understand the universe, he requires the atomicity of "a world [with] real individuals, each with a distinct identity and a definite character" (p. 520). On the other hand, Rosenthal (1996) states, "in our experience we normally discriminate not a single actual entity but a nexus of them united by their prehensions" (p. 543), and further, that the experience of this nexus is that of enduring defining characteristics persisting through ongoing change, so that continuity occurs through the potentialities of change. Whitehead (1978) is clear on this point right up to his insertion of the word 'perhaps,' stating:

> Owing to the vagueness of our conscious analysis of complex feelings, *perhaps* (italics added) we never consciously discriminate one simple physical feeling in isolation. But all our physical relationships are made up of such simple physical feelings, as their atomic bricks. (p. 237)

As Hosinski (1993) has pointed out, though it does not always appear the case, Whitehead's philosophy is grounded in ordinary human experience, as this "is both the source and the proving ground of philosophy" (p. 33) and the place to begin our exploration of the structures of experience and the constitution of actuality; therefore, let us return to the concrete matters of the traffic in Ho Chi Minh City (HCMC). Phúc, a professional truck driver in HCMC, gathered video footage of traffic from a GoPro camera mounted on the dashboard of his truck. This footage was viewed later in a viewing session, in which Phúc was present, as well as another traffic user, the present author as lead researcher, and a research assistant as interlocutor and translator. In one scene, we see a small dump truck turn out in front of his truck from a smaller feeder road, onto the larger road that Phúc was driving along. Although the dump truck is smaller than Phúc's truck and was coming from a smaller road onto a larger road (and hence should have given way), the dump truck continued out, seemingly oblivious to the presence of other vehicles, creating a situation, whereby Phúc had to take immediate action in order to avoid collision.

When viewing the video, Phúc explained the event:

> I have a *feeling* that this vehicle would come out and I am pretty sure about that as it won't let me go first. . . . For assessment, I know it is a dump truck and there is less chance for them to give way for me. In Vietnam, there are types of vehicles that force me to let them go first. It's in our *subconscious* [emphasis added], not because of the law . . . with my experience of more than 10 years of driving, I should let it go first to avoid crashes . . . . If I encounter this situation for 10 times, the vehicle would come out in all the 10.

In this series of events, there are many objects. The small dump truck, for example, is an object that exists as multiplicities, each of which arise through differences in relations: as a pure abstract representation of 'that kind of truck,' which, from experience, never gives way, and also as a felt presence, an embodied encounter of movement and perceived intention. In his narrative, Phúc alludes to a dual system of perception, of feelings, on the one hand, and abstractions remembered from past events on the other, as well as an acknowledgement that much of this knowledge is tacit, and that he is acting, to a large extent, through his subconscious. In this situation, the influence of the past is crucial and that influence is clearly being *felt* and, in a sense, 'relived.'

According to Merleau-Ponty (2005), to perceive is not to project memories or to bring into the present self-subsistent images of the past, but to be, in his words,

> Thrust deeply into the horizon of the past and take apart step by step the interlocked perspectives until the experiences which it epitomizes are *as if relived* [emphasis added] in their temporal setting. To perceive is not to remember. (p. 67)

The past, then, when it is reactivated by events in the present, is less like a memento, such as a postcard or seashell, picked up on a long-ago stroll along the beach, now sitting, disembodied on the desk in our study, but more like a stroll along that very same holiday beach, alive with its original affective intensities. When aspects of contemporaneous events accord with the past, we are once more immersed in a "field" or "horizon" (Merleau-Ponty, 2005, p. 66), driving along those same atmospheric corridors, where things retain the same kinds of deeply felt presence.

Furthermore, the virtuality of experience populates such reanimations from the past with its own kinds of objects, not necessarily physical, but yet creating their own temporalities and spatialities. Huy, who is a young Vietnamese man in his early 20s works full time delivering drinking water in large 20-litre bottles by motorcycle. There are two quite remarkable things about Huy's job, both of which, though remarkable, are extremely common practice in the HCMC traffic system. First, his traffic video, taken using a GoPro chest harness whilst riding, shows a dilapidated old motorbike, the front of which is held together with packaging tape that bears the name of a water company and tied up with a 'stretchy' cable (see Figure 6.1). When I pointed out the state of the bike, he told me that the turning lights and speedometer do not work, and so he uses his hands as turning signals instead. The second remarkable thing is that on this dilapidated old bike, he often has to carry eight or nine water bottles, weighing up to 180 kilograms in total.

This is a small bike and Huy is of small build, and even when he explains just how the bottles are stacked on the bike, it is hard to imagine how difficult riding would be with such a weight and with such limited space for the rider.

Huy, together with Phúc the truck driver, and two researchers, watched a 10-second traffic sequence of Huy's video, and, following this sequence, Huy was then asked to recall two different things that were present in the video (I was curious about the things that 'pop' out as memorable for someone who spends so much time on a motorcycle in traffic). Of course, the watching of a traffic video really bears little resemblance to actually riding a motorcycle in traffic delivering water bottles, especially in terms of affect, embodied

Figure 6.1 Screenshot from video footage taken from Huy's motorbike, on route to deliver water, 2016.

awarenesses, or subconscious processes of thought, but nevertheless, video can act as a stimulus leading back into these original events and processes. Huy pointed out a location on the street corner and described for us a scene involving four or five pet dogs that are fed out on the road. Not recalling seeing any dogs in the video, I then asked him to rewind the video, so that he could point them out to me. It turned out that the dogs were not actually present in the video at all, though they remained as a presence for him, meaningfully related to the house we saw in the video. Huy explained that the dogs are let out onto the road when they are being fed and he has almost hit them in the past. What is interesting about this response is that Huy was asked to recall something *actually present* in the scene, and yet he described a scene populated with virtual objects imported into the present from the past.

Just to make sure there was no confusion, I reminded Huy that the things he should try to point out in the video should be actually present in the scene. We then watched the same 10-second sequence and I asked him to recall something more. Huy then pointed out a house under construction on the left side. When I asked him if there was anything remarkable about this site, he described almost hitting a truck that was reversing out from the site into the street. Again, as with the scene involving the dogs, this was a scene, though stimulated by objects that were actually present in the video (the house and the construction site), the experience was relived with the aid of virtual characters imported from the past.

We can see through Huy's narrative, the open nature of events, how they flow back into the past and how the things that are physically present, open out fields and horizons of affective experiences that retain their aboriginal qualities. This concrete example highlights an aspect of Whitehead's metaphysical scheme that remains problematic for scholars: how something ontologically divided and discrete (actual entities) can simultaneously flow with experiential continuity (Rosenthal, 1996). The ontological structure of Whitehead's process philosophy is atomistic, and so actual entities form a discrete temporal serial procession, like dominoes one after the other (Whitehead, 1978). Whilst these individual building blocks are monads, discrete, and individual, they are open, in that they connect and implicate into one another. As we all know, when two dominos are placed end-to-end, no matter that they connect by their matching symbols, they can still never be a seamless continuity of one domino. This aspect of Whitehead's metaphysics remains a point of contention (Rosenthal, 1996); how can a reality built up of individual discrete 'drops' of experience be experienced as a seamless continuum, more like Bergson's notion of *duration*, rather than a bumpy ride across experiential gaps? Massumi's (2008) answer to this is that whilst experience can come in 'drops,' it also comes with more "oceanic" feelings because "every thing appears in a situation, along with others" (p. 12) and the situation itself has ripples of potential that flow out from it, drops becoming ripples, becoming flows, "with crests and troughs," "surf-breaks," and "gentle lappings" (p. 12).

Whitehead's monads are open and interdependent because "an event has to do with all that there is, and, in particular with all other events" (Whitehead, 1948, p. 105); events are the rock pools that contain, in varying degrees, oceans of past oceans. Whitehead's actual entities take on aspects of other events from the settled past by way of their *prehensions*, which are their particular forms of relations, the term 'prehension,' originating from the Latin, meaning to seize or grasp (Hosinski, 1993). As part of their process of becoming – the process Whitehead calls *concrescence* – actual entities take on aspects of other actual entities in the form of felt intensities or "feeling-tones" (Whitehead, 1978, p. 120), gradually building up more complex concrete events from which emerge the more abstract things in the world. So, whilst actual entities are not divisible into other contemporary actual entities, they are divisible into their prehensions or particular relations (Whitehead, 1978), and so the key to understanding how the experiential world is put together is by way of relations, rather than *relata*, process, rather than product.

Prehensions in concrescing actual entities are felt as feeling-tones, flowing between conscious and unconscious cognition, involving subjects by way of "emotion, and purpose, and valuation, and causation" (Whitehead, 1978, p. 19). The implication of this is that subjects are seen as being immersed in a world of felt intensities, and these are fundamental to the emergence of the trajectories of 'decisions,' aims, and purposes on the part of subjects. Whilst prehensions, as constitutive relationalities, are not necessarily consciously perceived or cognitized, they are still present in awareness as felt intensities, and Whitehead (1978) maintains, "every entity is felt by some actual entity" (p. 41). The theory of prehensions is, therefore, a theory of feelings at its most fundamental atomic level.

Whitehead's phases of concrescence, like the chrysalis that demarcates the different forms of caterpillar and butterfly, remains a *duree*-like process, an inaccessible blackbox, a Deleuzian fold, a "thick present" (Nail, 2019, p. 372), and even the most seemingly mundane event is like the *TARDIS* in *Dr Who*, deceptively simple, but vastly interconnected. In practice, any traffic event is continuously experienced, and, as a fold, it is like the inside of the outside (Deleuze, as cited in Halewood, 2005, p. 74), whereby even one single atomic actual entity has the relational influence of every other entity in the universe, once we allow for "degrees of relevance, and for negligible relevance" (Sherburne, 1981, p. 24). Actualities emerge both through positively incorporating the feelings of past events, or rejecting them, interdependently, within the greater web of other events.

Whitehead (1978) describes actual entities as the "final real things of which the world is made up" and furthermore, "there is no going behind [them] to find anything more real" (p. 18). Whitehead is able to make this assertion of 'realness' or, more accurately, fundamentalness, based on his description that actual entities, in forming the fundamental *building block-becomings* of the universe, remain divided from each other, but are not themselves divisible into anything more fundamental (Whitehead, 1978); in other words, they are *becomings* that are indivisible into further *becomings*. The idea that each singular actual entity is composed of individual character is suggested in Whitehead's description of these most basic of entities as "real, individual, and particular" (Whitehead, 1978, p. 20). This statement pertaining to individuality at the most fundamental level of actuality also implies that the world is built up, not from universals, as a more classical Newtonian view would suggest, but from particularities, the particular being central to Whitehead's scheme. Therefore, whilst abstractions (such as universals) can and do arise from more concrete things, the opposite does not hold true, in that the world cannot be built up from abstractions (which are also 'real' but in a different sense), a move that would fall prey to the fallacy of misplaced concreteness (Whitehead, 1978). Whilst actual occasions or entities are 'real,' even if they exist in a realm below the threshold of conscious perception and sensory awareness, "whatever things there are in any sense of 'existence,' are derived by abstraction from actual occasions" (Whitehead, 1978, p. 73). The implication of this is that consciousness always begins in the realm of unconscious affective experience, rather than the other way around.

Whitehead's writings suggest that these unique, 'one-off' occurrences or becomings, known as actual entities, exist not only beyond conscious perception but also beyond the temporal realm. Existing beyond temporality (yet *constituting* temporalities through this very process), any such singular events, in themselves, can never be 'known' or recognized because they lack the quality of endurance (they pop out as soon as they pop in) and therefore cannot be compared to any other contemporaneous actual occasion *in* time; this is why Whitehead (1920) says of an event, "when it is gone, it is gone" (p. 169). However, if the world only consisted of actual entities, swiftly becoming and then perishing one after the other, and if all things remained in this kind of constant flux, there could be no knowledge because, as Hooper (1942) states, any statement about things would be instantly changed to something else in the instance of its utterance. Experience tells us that this is not the case, and our ability to compare and to differentiate a 'now' from a 'then' and a 'here' from a 'there' occurs through endurances and temporalities that allow for experiences of contrast and comparison that 'there it is again' feeling (Stengers, 2011). As Whitehead (1920) says, we can "observe another event of analogous character [even while] the actual chunk of the life of nature – the actual entity – is inseparable from its unique occurrence" (p. 169), and this ability of recognition is essential to meaningful continuity.

These various statements thrown about in the previous few paragraphs require more rigorous and closer inspection, and it should also be noted that Whitehead's use of the terms 'object' and 'event' in *The Concept of Nature* does not align with his use of the terms 'actual occasions,' 'actual entities,' and 'eternal objects' in the later work *Process and Reality*, in which Whitehead employs the term 'actual occasions' in favour of the earlier technical understanding of 'event.' This book uses both sets of terms, and the differences between these terms will hopefully become more apparent over the course of the book, but, in brief, the term 'object' is used here to refer to what Whitehead (1978) describes in *Process and Reality* as "a genetic character inherited through a historic route of actual occasions" (p. 109).

Whitehead does not sharply circumscribe beings through criteria that strictly corral different kinds of existence, such as between sentient and non-sentient beings, and appears to acknowledge the role of the observer in drawing such conclusions, as implied in the following statement:

> We must remember the extreme generality of the notion of an enduring object – a genetic character inherited through a historic route of actual occasions. Some kinds of enduring objects form material bodies, others do not. But just as the difference between living and non-living occasions is not sharp, but more or less, so the distinction between an enduring object which is an atomic material body and one which is not is again more or less. Thus the question as to whether to call an enduring object a transition of matter or of character is very much a verbal question as to where you draw the line between the various properties. (1978, p. 109)

Whitehead's creativity also allows for situations that seem paradoxical, from which possibilities for confusion can arise. I earlier stated that actual entities exist beyond the temporal realm, which appears a strange statement for a book that proclaims a phenomenological approach. Certainly, the phenomenological experience of traffic is one of many different temporalities, all stratifying each other, each vying for conscious attention and also drifting surreptitiously through the back door of consciousness, so to speak, in vague and unconscious awareness: the pressure of an unanswered text message sitting in the inbox on the phone in my pocket while I ride, the LCD countdown timers that are attached to traffic lights, the physical experience of movement and space, the actions of other traffic entities (both human and non-human), the *substantializing* effects of ideas, beliefs, and immaterial objects, and the emergent effects and affects of various complexus, infrastructures, systems, and multiplicities. But does this temporal-based experience automatically exclude the possibility, from a speculative metaphysics perspective, of the existence of an atemporal realm?

Whitehead (1978) describes an actual entity as a "drop of experience" (p. 18) or "unit of complex feeling" (p. 80), and one would think that such experience would involve some kind of awareness and perception (conscious or otherwise), and that awareness must involve an experience of temporality (Landgrebe, 1949), and also, as Reese (1965) points out, the experience of order. Rather than time being merely present in consciousness, Landgrebe (1949) says, "time produces itself in the 'performances' of consciousness" (p. 204), and what we mean by 'time' is the product of these performances and dependent on them.

What of the experience of riding my motorcycle when my mind wanders, before my conscious awareness, once more, alights upon something more fixed or settled? Of course, even in the midst of such 'flights of fancy,' I may still feel the affectivity of *relations* in the traffic, and the very real feelings, as William James' (1918) radical empiricism explores, of the kinds of conjunctions and prepositions that are reflected in language. James (1918) says that it makes as much sense for us to speak of the feeling, for example, of 'and,' the feeling of 'but,' the feeling of 'if,' and the feeling of 'by,' as much as it does to speak of the feeling of 'blue' or of 'cold.' In the same way, I may feel a 'being-with' amongst other entities in traffic, a synergy of being in synch, or the opposite, of entities with conflicting tendencies, of things moving against me.

James (1918) also speaks of the stream of consciousness as being like a bird: "an alternation of flights and perchings" or resting places, where 'flights' represent the "transitive parts" and 'perchings,' the "substantive parts" (p. 243), a rhythm also represented in language, where each word functions as a perching, a decision, and a stepping-stone. The question then becomes, what is the nature of the topology of the 'flights' – whose complexity and vagueness defy reductionism – and the relationship between the transitive and the substantive? James (1918) describes the difficulties in the dissection and analysis of those moments of life that reside in the very blurred edges of awareness, seemingly outside of time:

> It is very difficult, introspectively, to see the transitive parts for what they really are. If they are but flights to a conclusion, stopping them to look at them before the conclusion is reached is really annihilating them. Whilst if we wait till the conclusion be reached, it so exceeds them in vigor and stability that it quite eclipses and swallows them up in its glare. (pp. 243–244)

Rosenthal (1996) says that it is only through the discrete seriality of the birthing and perishing of actual entities that we get tenses and, therefore, time, but that the processes that create time occur in a non-temporal realm. In Rosenthal's words (1996), "apart from the succession of actual entities time does not exist, for this succession constitutes 'the becoming of time.' Actual entities are not 'in time' but rather their atomic succession gives rise to time" (p. 543). This means, as Rosenthal says, "actual occasions, then, are temporal atoms or temporal building blocks, but are not themselves temporal. Each one is an indivisible epoch having no internal temporal phases" (p. 543). In agreement, Chappell cites Whitehead's statement that the act of becoming that results in an actual occasion "is not extensive" (Whitehead, cited in Chappell, 1961, p. 518), meaning, not temporally extended (not in time), and Whitehead (1978) himself describes the process of concrescence that results in an actual entity thus: "this genetic passage from phase to phase is not in physical time" (PR, p. 283). There appears some consensus, therefore, that the processes that give rise to actual occasions occur beyond the temporal realm, but what of the products of these processes, the atomic actual occasions?

It was earlier stated that actual entities or occasions are simultaneously both a process and a 'thing,' a statement that appears at odds with itself, especially in light of the fact that the processes are said to occur beyond the temporal realm, whilst the products of such processes are temporal and reside in the temporal realm. It seems that there are two interrelated issues at stake here: first, by what criteria or at what point are we able to distinguish between an actual entity as open process, unresolved, and fluid and an actual entity as a product or a settled fact, or what Whitehead (1978) refers to as the *satisfaction* (or are they, in effect, one in the same, constitutive of each other?), and, second, if any 'single' actual entity is not divisible into anything more fundamental, as Whitehead states, are we able to have an experience that allows us an awareness of its singular individual character? Even scholars who maintain that the processes that give rise to time occur outside of time often employ language embedded in notions of time to describe those processes. For example, Rosenthal (1996), who says, "the moment of becoming of an actual entity is also its moment of perishing" (p. 542) or Whitehead (1978) himself, who employs terms such as "succession," "phases," and "antecedent" (p. 26).

Whitehead (1978) suggests, "there are two distinct ways of 'dividing' the satisfaction of an actual entity into component feelings: genetically and co-ordinately. Genetic division is division of the concrescence; coordinate division is division of the concrete" (p. 283), and it is this genetic passage, whose phases (a problematic term, given that 'phases' suggest a linear temporal sequence) are not in physical time. This seems to imply that it depends on the mode of analysis or the mode of division as to the constitution and definition of an actual entity, and though Whitehead maintains a contrast "between the *product* and the *process* of becoming, whether microscopic or macroscopic" (Chappell, 1961, p. 518), he also employs a double usage of the word 'time' (Chappell, 1961). This distinction is expressed by Whitehead (1978) in the statements: "physical time expresses some features of the growth, but not the growth of the features" (p. 283) and "there is a becoming of continuity, but no continuity of becoming" (p. 35). Chappell (1961) says that Whitehead's whole rationale is that something must become from the process of becoming and the 'becomed,' if you like, is something with temporal extension and is critical of the fact that Whitehead's epochal theory of time rests on the notion of a becoming that does not extend through a certain period of time, implying a process that takes no time to occur. Hosinski (1993) says, "Whitehead holds that the concrescence of an actual entity happens 'all at once' in a 'quantum' of time (a small fraction of a second)" (pp. 46–47) and that Whitehead speaks of the processes as being in 'phases' only as a way of logical analysis and not to suggest that they actually occur as phases in any kind of linear process. In light of the fact that the etymology of the term 'concrescence' describes a growing together, congealing, or hardening process and that Whitehead, in using this term, wants to understand the *process* through which a moment of experience arises, grows together and forms, and how the 'fluidity' of becoming 'hardens' into being (Hosinski, 1993), we only seem to venture further into murkier waters. The problem seems to me more to do with persisting tendencies towards dissection and reductionism and that we are trying to dissect a phenomenon that resists dissection.

Other scholars also differentiate between the act of becoming and the product of actual entity, such as Pomeroy (2004), who states, "the act of becoming of any actual entity is non-temporal" (p. 30) and differentiates between the entity as subject, which is the non-temporal act of becoming, and the entity as *superject*, which is the temporal extension of the entity in other acts of becoming. The confusion arises because concrescence – the act of becoming of actual entities – consists of prehensions, and prehensions, which are forms of relatedness, have a subject, but the subject cannot be antecedent to the prehensions, nor subsequent to the concrescence, because the subject only arises with process, and there cannot be prehensions without a subject (Baxter & Taylor, 1978); in other words, we emerge simultaneously with world. It should be noted that Whitehead (1978) separates the processes involved in the becoming of an actual entity into two different species: "concrescence" or final causation and "transition" or efficient causation (Hosinski, 1993). According to Nobo (1986), "transition is the creative process providing the conditions really governing the attainment of an actual occasion, whereas concrescence is the creative process providing the ends actually attained by an actual occasion" (p. 139). In other words, transition involves the process of efficient causation from one actual entity to another, and concrescence involves final causation, expressing "how each actual entity individually becomes itself" (Hosinski, 1993, p. 96). These two processes occur simultaneously and help explain continuity amid atomicity, as the process involves the actual entity not only taking on actualities of the past but also anticipating and objectifying the future into the present (Nobo, 1986).

As Shaviro (2009) notes, Whitehead "privileges feeling over understanding, and offers an account of experience that is affective rather than cognitive" (p. 56), describing a universe where subjects make 'decisions' primarily through interactions characterized by feelings, so much so, that an actual entity is *inseparable* from the feelings that are affecting it; as Whitehead (1978) writes, "The feelings are what they are in order that their subject may be what it is" (p. 222). For Whitehead, even if we focus our attention on the *sensa* (sense perception), the emotional significance of this *sensa* is still its main characteristic. The idea that we perceive first and then respond emotionally second is an illusion and presents things in a mistaken order: "Perception is first of all a matter of being affected bodily" (Shaviro, 2009, p. 56). Prehensions, experienced as felt propositions, are what Whitehead calls "lures for feelings" (1978, p. 280), and this aesthetic dimension is central to how new *concrescing* actual entities take on characteristics and aspects of the settled past. In other words, we feel the feelings in the objective data that arise from past events, and this notion – that felt embodied awarenesses are of central importance to decision-making – is of particular relevance in the development of a phenomenology of a traffic system such as in HCMC, which is so affective, rhythmic, and emotion-laden.

However, as is the case with much of the vocabulary of Whitehead's scheme, care should be taken to understand the unique applications of what are sometimes more general terms. Whitehead employs the term 'feeling,' not as an emotional term (Hosinski, 1993), but more as a technical term. However, in my opinion, Whitehead's (1978) description of our primordial way of being in the world, as inspired by *feeling-tones*, suggests that the term 'feeling' still retains content similar to what is often referred to by the term 'affect.' However, not all scholars would agree, and one writer suggests that Whitehead does not even use the term 'feeling' to denote *awareness*, but to simply refer to a process of inclusion (Johnson, 1945).

If actual entities, which are the building blocks of reality, lack the quality of endurance, perishing as soon as they are born, what is it then that carries forth and endures, allowing meaning and knowledge to emerge from a world in complex and constant flux? The answer, according to Whitehead (1920), is 'objects.' However, when we consider the event as the most fundamental thing in nature (as opposed to substance), all commonsense notions of material objects must be put aside (Stebbing, 1926). What we have become accustomed to thinking of as stable, concrete, and independent – objects – are, for Whitehead, a more ephemeral emergence of recognition of character. Rather than start with substances that have predicates, which then give us events, Whitehead upturns the whole landscape and begins with relational events as the fundamental reality, from which abstractions such as time, space, character, and objects then emerge. Therefore, it is not individual events that endure, but the subjective forms and the particular modes or patterns that arise from out of events – the 'how' of the relations – and, for this reason, it is *relations* that should be thought of as 'substantial' rather than physical objects, which are always undergoing constant change. Therefore, an actual entity is both a 'how' and a 'what,' and so *how* it becomes constitutes *what* it is (Hooper, 1941; Shaviro, 2009).

This thing that endures and that is recognized from out of the swift comings and goings of micro-events or micro-becomings (actual entities or actual occasions) is *character*, and, as Whitehead (1920) says, "things which we thus recognize I call objects" (p. 169). Therefore, it is partly through our experiences with 'objects' in their various guises that we come to experience time and affectively experience the character of events, or, to put it another way, objects *are* the character that flows through events. The fundamental (universal) quality of objects, then, is not that they *have* qualities, but that they have the quality of endurance, recognizability, and of being "the elements in nature which can 'be again'" (Whitehead, 1920, p. 144). Rather than having innate qualities, such as roughness or smoothness, they have the quality of 'being again.' Though Whitehead's ontology *does* deny objects their Cartesian substance-like independence, it does *not* deny them their qualities of endurance, and it is only through the enduring, persistent – even if ephemeral – presence of objects that we are able to have meaning-filled awarenesses of events. To deny the notion of a substance ontology, that nature consists of a substratum of enduring and unchanging substance, is not to deny the idea that all of experience depends upon a substratum of continuing and enduring forms and the maintenance of such structures (Hosinski, 1993).

Therefore, objects may be thought of as the enduring ghosts of reality. They are caricatures, opaque facades, and emergent holographic representations, appearing mirage-like, as a nexus of shifting relations, existing at the shimmering edges of the things-in-themselves. What we know as a 'car,' or any other object, therefore, emerges from a series of events as a kind of ghostly coalescence of character, a relational phenomenon or collective event, of which, I am an entangled, inseparable part. Objects are, therefore, emergent abstractions, in the same way that time and space are emergent abstractions, existing as an enduring stream of character that persists through a "complex of passing events" (Whitehead, 1920, p. 166), and they are what give events their character, as clearly stated from Whitehead (1920): "an object is situated in those events or in that stream of events of which it expresses the character" (p. 169), and so objects are the "permanences" that events "body forth" (Whitehead, 1920, p. 144).

Our awareness of the character of events is possible because objects are defined by two fundamental dimensions: their constitution as a genetically inherited character that endures through a temporal series of events and as a marker of location in an entire structure of other events or existential system of meaning (Whitehead, 1920). In regard to these dimensions, Whitehead (1920) says:

> The character of an event is nothing but the objects which are ingredient in it and the ways in which those objects make their ingression into the event. Thus the theory of objects is the theory of the comparison of events. (p. 143)

From this, we may surmise the existence of three interrelated dimensions that are crucial for experience and, hence, for any development towards a phenomenology of traffic: character, objects, and recognition of object/events in time. Without the ability to recognize, to have that aha-there-it-is-again experience, we would be forever lost in shimmering, meaningless mobility, rather than the meaning-constituted web or ecology that is my being or my traffic world. Therefore, any happening in traffic may be described by recourse to three fundamental dimensions: *objects*, *events*, and *character*, each of which must be seen as one side of a three-sided coin. What endures across time, hence, *constituting* our experience of time, are not events, but *prehensions*: relational patterns, character, modes, and forms, enduring not as fixed settled facts, but as *potentiality*; as Whitehead (1978) clearly states, "it is not 'substance,' which is permanent, but 'form'" (p. 29). These patterns or forms are what we recognize and abstract into our 'objects,' and it is only through objects, as ingredients in events, that "a character of an event can be recognized" (Whitehead, 1920, p. 169), and so we recognize the nexus as being united by their prehensions rather than the individual actual entities that are constantly popping in and out of existence (Rosenthal, 1996). Though actual entities themselves perish, forms and relations endure as immortally objective data, so that future concrescing actual entities can take in aspects of the settled past by *prehending* these relations, like a runner passing the baton to the next (Shaviro, 2009). Importantly, though prehensions are simple physical feelings of the datum of the past, a past of objectively immortal, settled facts, these 'fixed' entities are felt from the standpoint of the newly concrescing occasion, thereby constituting a new kind of relation that gives the past a quality of unresolved openness (Stengers, 2011).

If this is beginning to sound confusing, I offer the following: Whitehead is as frustrating to read, as he is creative and ingenious. Sherburne's (1981) statement that Whitehead "is at no point excessively clear on these matters" (p. 66) appears as a familiar theme when reading Whitehead, and, in fact, the editor's preface (by Griffin and Sherburne) in Whitehead's magnum opus, *Process and Reality*, starts out with the note: "the work itself is highly technical and far from easy to understand" (1978, p. v). Urban (1938), who is a critic of Whitehead, wrote that *Process and Reality* "proved to be almost the most unintelligible essay in philosophy ever written" (p. 617), and Sherburne (1981), the author of the book, *A Key to Whitehead's Process and Reality*, calls it "an extremely difficult book" (p. 1).

Given that Whitehead (1978) says that reasons are always to be found by way of reference to actual entities in the concrete world, the best way through the Whitehead 'jungle' is by way of a real-world story. A recent on-line video (Saigoneer, 2017) shows a large green government bus in HCMC, mounting the curb and travelling along the sidewalk, in order to drive around a traffic jam, a move that is obviously illegal and dangerous. Such a manoeuvre would appear commonplace in HCMC if the vehicle were a motorcycle, but the sight of a bus performing the same kind of action is, in my experience, extremely rare. Having said that, other than the pedestrians at the sidewalk café who were videoing the bus, the event barely appears to register with other traffic users in the video, and none of the motorcyclists on the street even turn their heads towards the bus. From the perspective of a viewer watching the video, or even from a pedestrian bystander involved in the event, this mount-the-sidewalk strategy is instantly recognizable as a characteristic of HCMC traffic. However, though it is recognizable, it is a characteristic never usually seen in a bus, is definitely not *bus-like*, which is, no doubt, why it was videoed and uploaded to the Internet.

On one level, the mount-the-sidewalk strategy has an instrumental aspect to it and is based in affordances – the bus did so because it physically could – but there is more to affordances than mere physicality. There is also something relevant and inviting, the bus has been *lured* onto the sidewalk. The event is coloured or characterized with a particular atmosphere, a feeling that I have felt elsewhere, the recognition of something previously experienced. It has been infused with a particular character that is atmospheric, moulding the situation into a certain form and feel. In Whitehead's terms, something has been *ingressed* into the event and that something is an *eternal object*. I will call this eternal object that has been ingressed into the bus-event, *motorcycleness*, or *fluid-motorcycleness*. Whatever its name, it is a dominant and recognizable characteristic in the traffic in HCMC, a kind of boundary-crossing defiant resilience, at one with the affordances of the motorcycle, but also present in many other domains of Vietnamese culture. *Motorcycleness* is a familiar and recognizable form in HCMC traffic, enduring through past events and then *ingressing* into this mount-the-sidewalk-bus-event just described. For my part, when watching the video, I had a feeling of 'there it is again,' part cognitive concept and part feeling – a *prehension*. What Whitehead calls 'feeling-tones' are inseparable – or antecedent to – mental concepts, in the tradition of James (1996), that thoughts and things do not, initially, exist in dualistic realities. Therefore, before objects might become conditions for cognitive understanding and mental abstractions, they are, more primordially, *subjective forms of feelings* (Shaviro, 2009). As Whitehead says, "perception, in this primary sense, is perception of the settled world in the past as constituted by its feeling-tones, and as efficacious by reason of those feeling-tones" (Sherburne, 1981, p. 100). Such a process involves dimensions below the level of conscious abstract thought and undetectable through traditional empirical means, and, for this reason, the phenomenology of the becoming of the subject can only be a "phenomenology of the inapparent" (Dastur, 2000, p. 181).

Manning (2009) says that part of the experience of prehending 'chair,' for example, is that we *feel* "sitability" (p. 7) as a modality of chair-expression. The chair is thereby experienced as a fusion of the mental feeling and the physical feeling which is its subjective form, and so the experience we have is more "chair-as-sitability," a vague feeling that registers below the range of sensory awareness. The reason we do not consciously realize that we experience chair and sitability together, as both physical and mental 'feelings,' is because they are so intertwined, forming one and the same experience. However, in the story above, I had a feeling of bus-as-motorcycleness, a combination not usually experienced as one. This character of *motorcycleness*, like *sitability*, is more than just a physical affordance in the materiality of the bus, but rather, an eternal form or mode that endures in the traffic in HCMC. It is a subjective form and what Whitehead (1978) calls a "lure for feeling" (p. 185). *Motorcycleness* and *sitability* are both examples of *eternal objects* that may be ingressed into an event and, therefore, are no different in nature from a sensory eternal object such as *redness*. *Redness*, as an eternal object, is timeless and non-spatial in the same way that *chairness* and *motorcycleness* are, and whilst redness might be more abstract, it is not different *in nature* from other kinds of character, different in degree, but not in kind (Stebbing, 1926).

Motorcycleness is a subjective form or mode of being that arose from out of the 'snows' of yesteryear, and in answer to the question, 'where are these snows of yesteryear?,' the Whitehead answer leads us into the eternal realm and, in doing so, builds bridges between the actual and the eternal. Whitehead conceives of only two fundamentally different kinds of entities in existence in the universe: those that deal in what is possible and can tell us something about the forms of possibilities, termed *eternal objects*, and those entities that deal in concrete fact, termed *actual entities* or *actual occasions* (Hosinski, 1993). Eternal objects are timeless potentials, similar to Platonic forms, existing as pure forms outside of the temporal and spatial realm (hence the term 'eternal'). Whitehead (1978) defines them as "pure potentials for the specific determination of fact, or forms of definiteness" (p. 22). In other words, an eternal object is an inexhaustible entity of potential that ingresses into spatial and temporal outcomes in the concrete world, providing potentiality and possibilities for novelty in manifested actualizations of settled facts.

Whilst eternal objects are not actual, as Shaviro (2009) says, "they haunt the actual" (pp. 42–43). For this reason, Whitehead says that eternal objects "tell no tales" (Sherburne, 1981, p. 60) as to their *ingressions*, meaning that they are only known to us as eternal forms by way of logical concepts outside of sense awareness and can only be apprehended by senses when actualized in a particular concrete instance. For this reason, they are not *substantive*, in that actualized fact can never be built from eternal universal forms or templates, because the abstract can never explain what is concrete (Stengers, 2011). Rather, it is the inverse: Whitehead's ontological principle states that the reasons for things are always to be found in the particularities of definite actual entities (1978, p. 19).

Eternal objects can be sensory qualities such as colours, but they can also be all kinds of affects, emotions, contrasts, patterns, numbers, shapes, and tactile qualities such as smoothness (Shaviro, 2007b). It is eternal objects, therefore, that are responsible for feelings that come to be categorized in actuality (though no such categorizations exist in eternality) as intensities, emotions, adversions or aversions, and pleasure and pain (Whitehead, 1978). Eternal objects can be simple, such as sense data, or complex, such as affect, emotions, patterns, contrasts, or, in fact, all of the different ways in which eternal objects might experience relatedness (Shaviro, 2009, p. 38).

For example, 'redness' exists as an entity outside of sensory awareness and outside of time and space. We only come to have an experience of redness as part of particular experiential concrete events in the actual world in which this eternal object is felt. As an eternal object, *redness* contains all the infinite possibilities of actual manifestations and experiential encounters, and these could be as varied as the experience of a polished red Lamborghini as it whizzes by with a roar of its engine, the soft red plushness of Salvador Dali's 'Mae West' sofa, or a dusty red stop sign on a country road. As an eternal object, *motorcycleness*, likewise, could be ingressed into any number of events, as part of traffic events or not, thereby infusing an event with character.

The question arises as to whether HCMC traffic is more prone to the ingression of certain kinds of eternal objects over others. It is said that no new eternal objects ever arise (Root, 1953) (somehow they have just always been around), but, as Hooper (1942) says, the number of possible combinations of eternal objects is "large, if not infinite" (p. 59). The *limitless* potential provided by these pure forms is then *limited* in actuality by prehensions, which are concrete facts of relatedness (Rosenthal, 1996). As part of the process of concrescence, eternal objects may be included and incorporated into event outcomes or rejected, but in either case, they allow us to say what a thing is. As an eternal object, *redness*, *motorcycleness*, or *carness* is neither true nor false, but only becomes so, once it is manifested in actuality. Therefore, it is the nature of objects that they are always multiple, always more than a single appearance, exuding an excess that extends out beyond any present incarnation, beyond the discernible, bridging the actual and the eternal, the definite and the indefinite, and the past and the future. The emergence of an object then functions as a signpost, bringing into view whole relational complexes of what *is* and what is *not*, what *was* and what *could be*. Without the aid of objects, emerging as meaningful and recognizable abstract markers, we would remain in what Stengers calls "pure, shimmering, mobility" (Stengers, 2011, p. 74), paradoxically immobile, in a realm of unlimited potentiality, without 'decision,' closure, or order, and without aims, goals, relevance, or significance. Therefore, objects, in their semblance as abstractions, give closure to the unlimited potentiality of the possible. Objects bring order to the flux of events by cordoning off spatialities and temporalities through manifesting the infinite forms of the universe into particularities and giving us more solid and reliable "footholds" (Whitehead, 1920, p. 107) in reality.

Character, then, must be seen to be a relational construction, emerging only as an endurance passing via genetic inheritance through events (of which we have an awareness by way of objects). Therefore, any quality or attribute cannot be present in an atomistic instant in a singular momentary event known as an actual entity. This is not the case in Aristotle's universe, where the subject–predicate proposition 'this is red' requires only the existence of the moment of 'this' and of 'red' (Stebbing, 1926). In the traditional subject–predicate model, the world is imagined as divided into universal, unchanging, and enduring matter (a stone), on one side, and the experience of its attributes, a psychological addition (such as greyness), on the other. However, in Whitehead's process philosophy, objects are for events, what attributes or qualities are for substances in a substance-based ontology, and so events may be thought of as similar to substances, whilst objects (though not all objects) function similarly to attributes. For Whitehead, nothing can ever be truly independent, and so qualities can never belong independently to a substance (Stebbing, 1926). Therefore, the Aristotelian notion of a 'quality,' meaning a characteristic that is present at an instant, is not possible (Stebbing, 1926).

What does all this have to do with traffic? Urry (2004) describes 'automobility' as having a "specific character of domination" (p. 25), a statement that draws on an earlier Heidegger quotation that reads, "machinery unfolds a specific character of domination," itself, as cited by Sheller and Urry (2000, p. 737) in an earlier paper. The transference of this Heideggerian theme, from technology, in a general sense, onto, specifically, 'the system of automobility,' implies that there exists, not only a specific character but one that is universal and essential, proliferating, either, *as* 'automobility,' throughout the world, or as a result *of* 'automobility,' the two not being the same thing. Given the above discussion on Whitehead's notions concerning character, what are the implications of such a statement? If this universal character exists, does it emerge from persisting consistency in the nature of the relationalities involved, and, if so, do the same modes of relations persist and remain fixed, even as the system is imported and settles itself into other places? Should we begin our investigations into the nature of traffic based on the premise of such an abstraction, of automobility having a universal essential character, or begin from concrete events, working towards new abstractions of givenness?

Heidegger's views on technology have often been read as essentialist, though some scholars warn against such a reading (as cited in Borgmann, 2005). However, the question should be seen as more concerned with focusing on the forms of relations, rather than the components, and of retaining a fluid conception of essentialism, one based on particularities and multiplicities of appearances. In this way, we may begin with actual concrete events as reasons, rather than with universal abstractions that are static and timeless, and that do not necessarily arise from concrete actuality. Urry lists six universal components of the *system of automobility* that might contribute to this 'specific character,' including manufacturing, consumption, the complex of industries, such as oil companies, roadside hotels, etc., what he calls the "quasi-private" character of the car, environmental implications, and semiotic textural discourse around cars. The approach is infrastructural, but tending more towards the components or nodes that make up the structure of the system, rather than the *modes* of the relations themselves – nodes rather than modes. This is made clear in the following sentence: "the *complex* character of such systems *stems from the multiple fixities or moorings* [emphasis added] often on a substantial physical scale that enable the fluidities of liquid modernity" (Sheller & Urry, 2006, p. 210).

In focusing on the 'moorings,' any complex character might still evade us because it is an emergent outcome, located nowhere and everywhere, something that Urry, given his background in complexity, would have no doubt considered. Nevertheless, such an approach has difficulty in bridging eternality and actuality. Character, as we have just seen, is as much the result of the limitless potential of the eternal, manifesting into actuality, as well as the *presence of absence*, neither of which would be found in the 'moorings.' This strange behaviour from our bus in the above story is 'strange,' because it leads out towards the eternal realm, suggesting other alternatives. As *busness* was displaced – though still present in absence – in favour of the ingression of *motorcycleness*, a whole realm of potentiality comes into view. As Hooper (1942) says, "the understanding of actuality requires a reference to *ideality*, and this ideality is the realm of alternative suggestions [which] is the realm of eternal objects" (p. 52).

A network approach that attends predominantly to the concrete components of the system, overlaid with more representationalist semiotic readings, is not uncommon in the literature of the sociology of traffic. For example, much of the recent book, *Cars, Automobility and Development in Asia* (Hansen & Nielsen, 2017) – one of the few books that depicts the current states of contemporary urban traffic in Asian cities – often proceeds in a similar fashion, with some chapters drawing specifically on Urry's 'system of automobility.' In fact, Hansen (2017), who wrote a chapter about the traffic in Vietnam, following Urry, coined the term "system of moto-mobility" (p. 106), thereby adapting Urry's paradigm to Vietnam's unique motorcycle traffic. In line with the rest of the book, Hansen's chapter, entitled "Doi Moi on Two and Four Wheels: Capitalist Development and Motorised Mobility in Vietnam," outlines the role of the auto industries in the changing dynamics, consumption practices, and the symbolic values attached to different kinds of mobility practices. Hansen views the phenomenon as a socio-technical-material system and, therefore, explores how the components of the system, such as the corner motorcycle repair shops, all play their parts in the network, as contextualized within historical trajectories of change. Therefore, as is common in the sociological literature on traffic, the approach is representational, through its focus on the car as a symbol, and a 'networked' approach that generally prioritizes *relata* over relations.

The book's two main sections are devoted, first, to the complex that is the car industry in the selected Asian countries, and, second, to the consumption of the car, its value as a cultural sign, and its relationship with the various histories of the countries. This is not a criticism, as it is a productive approach in which a sociological study might well proceed, and given that the scope of the book is described as concerned with "policies, production forms, labour regimes, consumer aspirations and symbolism that are implicated in Asian automobilities" and the auto industries in Asian countries (Hansen & Nielsen, 2017, pp. 3–4), it is wholly appropriate.

Because traffic is considered a socio-technical complex system, frameworks for its analysis often reflect certain models of representation, such as technical schematics constituted by fixed elements, components, or nodes.

Adey (2006) describes how academic theorizations of movement often view mobility as something other than the norm due to a humanist focus on "fixity and boundedness of place and territory" (p. 77), because, he says, "power structures of domination are given fixed and immobile characteristics" (Adey, 2006, p. 77). The solution to this lay in an approach that begins with particularities, rather than with universal, timeless essentialism, and that focuses on the subjective relations, rather than the 'moorings,' viewing objects as ephemeral endurances of character and as processes, rather than fixed constructs.

References

Adey, P. (2006). If mobility is everything then it is nothing: Towards a relational politics of (im) mobilities. *Mobilities, 1*(1), 75–94.

Baxter, G. D., & Taylor, P. M. (1978). Burke's theory of consubstantiality and Whitehead's concept of concrescence. *Communications Monographs, 45*(2), 173–180.

Borgmann, A. (2005) Technology. In H. L. Dreyfus & M. A. Wrathall (Eds.), *A companion to Heidegger* (pp. 420–432). Malden, MA: Blackwell Publishing.

Brown, B. (2001). Thing theory. *Critical Inquiry, 28*(1), 1–22.

Chappell, V. C. (1961). Whitehead's theory of becoming. *The Journal of Philosophy, 58*(19), 516–528.

Dastur, F. (2000). Phenomenology of the event: Waiting and surprise. *Hypatia, 15*(4), 178–189.

Garland, W. J. (1982). Whitehead's theory of causal objectification. *Process Studies, 12*(3), 180–191.

Halewood, M. (2005). On Whitehead and Deleuze: The process of materiality. *Configurations, 13*(1), 57–76.

Hansen, A. (2017). Doi moi on two and four wheels: Capitalist development and motorised mobility in Vietnam. In A. Hansen & K. B. Nielsen (Eds.), *Cars, automobility and development in Asia: Wheels of change* (pp. 113–129). London, United Kingdom: Routledge.

Hansen, A., & Nielsen, K. B. (Eds.). (2017). *Cars, automobility and development in Asia: Wheels of change*. London, United Kingdom: Routledge.

Harman, G. (2007). On vicarious causation. In R. Mackay (Ed.), *Collapse II* (pp. 187–221). Falmouth, United Kingdom: Urbanomic Media Ltd.

Heidegger, M. (1996). *Being and time: A translation of sein und zeit* (J. Stambaugh, Trans.). New York: State University of New York Press.

Hooper, S. E. (1941). Whitehead's philosophy: Actual entities. *Philosophy, 16*(63), 285–305.

Hooper, S. E. (1942). Whitehead's philosophy: Eternal objects and God. *Philosophy, 17*(65), 47–68.

Hosinski, T. E. (1993). *Stubborn fact and creative advance: An introduction to the metaphysics of Alfred North Whitehead*. Lanham, MD: Rowman & Littlefield Publishers.

James, W. (1918). *The principles of psychology* (Vol. 1). New York, NY: Henry Holt and Company.

James, W. (1996). *Essays in radical empiricism*. Lincoln: University of Nebraska Press.

Johnson, A. H. (1945). Whitehead's theory of actual entities: Defence and criticism. *Philosophy of Science, 12*(4), 237–295.

Landgrebe, L. (1949). Phenomenology and metaphysics. *Philosophy and Phenomenological Research, 10*(2), 197–205.

Latour, B. (2008). What is the style of matters of concern. In *Two lectures in empirical philosophy.* Department of Philosophy of the University of Amsterdam, Amsterdam: Van Gorcum. Retrieved from http://www.bruno-latour.fr/sites/default/files/97-SPINOZA-GB.pdf

Manning, E. (2009). *Relationscapes: Movement, art, philosophy.* Cambridge, MA: MIT Press.

Massumi, B. (1995). The autonomy of affect. *Cultural Critique: The Politics of Systems and Environments, Part II, 31*, 83–109.

Massumi, B. (2008). The Thinking-feeling of what happens. *Inflexions,* 1, 1–40. Retrieved from http://www.inflexions.org/n1_The-Thinking-Feeling-of-What-Happens-by-Brian-Massumi.pdf

Merleau-Ponty, M. (2005). *Phenomenology of perception.* New York, NY: Routledge.

Nail, T. (2019). *Being and motion.* Oxford, United Kingdom: Oxford University Press.

Nobo, J. L. (1986). *Whitehead's metaphysics of extension and solidarity.* Albany: State University of New York Press.

Pomeroy, A. F. (2004). *Marx and Whitehead: Process, dialectics, and the critique of capitalism.* Albany: State University of New York Press.

Reese, W. L. (1965). Phenomenology and metaphysics. *The Review of Metaphysics, 19*(1), 103–114.

Root, V. M. (1953). Eternal objects, attributes, and relations in Whitehead's philosophy. *Philosophy and Phenomenological Research, 14*(2), 196–204.

Rosenthal, S. B. (1996). Continuity, contingency, and time: The divergent intuitions of Whitehead and pragmatism. *Transactions of the Charles S. Peirce Society, 32*(4), 542–567.

Saigoneer. (2017, June 5). Saigon bus driver disciplined for driving on the sidewalk. *Saigoneer.* Retrieved from https://saigoneer.com/saigon-news/10269-bus-driver-disciplined-for-driving-on-the-sidewalk

Shaviro, S. (2007a). Deleuze's encounter with Whitehead. Retrieved from http://www.shaviro.com/Othertexts/DeleuzeWhitehead.pdf

Shaviro, S. (2007b, May 13). Eternal objects (Blog post). Retrieved from http://www.shaviro.com/Blog/?p=578

Shaviro, S. (2009). *Without criteria: Kant, Whitehead, Deleuze, and aesthetics.* Cambridge, MA: MIT Press.

Sheller, M., & Urry, J. (2000). The city and the car. *International Journal of Urban and Regional Research, 24*(4), 737–757.

Sheller, M., & Urry, J. (2006). The new mobilities paradigm. *Environment and Planning A, 38*(2), 207–226.

Sherburne, D. W. (Ed.). (1981). A key to Whitehead's process and reality. Chicago, IL: Chicago Press, Macmillan.

Stebbing, L. S. (1926). Professor Whitehead's "perceptual object". *The Journal of Philosophy, 23*(8), 197–213.

Stengers, I. (2011). *Thinking with Whitehead: A free and wild creation of concepts.* Cambridge, MA: Harvard University Press.

Urban, W. M. (1938). Elements of unintelligibility in Whitehead's metaphysics. *The Journal of Philosophy, 35*(23), 617–637.

Urry, J. (2004). The 'system' of automobility. *Theory, Culture & Society, 21*(4–5), 25–39.

Whitehead, A. N. (1920). *The concept of nature: Tarner lectures delivered in Trinity College*, November 1919. Cambridge, United Kingdom: Cambridge University Press.

Whitehead, A. N. (1948). *Science and the modern world.* New York, NY: The New American Library of World Literature.

Whitehead, A. N. (1978). *Process and reality: An essay in cosmology (corrected edition).* New York, NY: The Free Press. Originally published in 1929 by Macmillan.

Whitehead, A. N. (1985). *Symbolism, its meaning and effect.* New York, NY: Fordham University Press, The Macmillan Company.

7 What is a car?

Plato's car

In many cities, cars are so ubiquitous and mundane an object as to form a kind of mobile infrastructure that almost recedes into the background, and it is probably for this very reason that the various impacts and effects of cars largely escaped the notice of academic researchers until surprisingly late in the day (Dant & Martin, 2001; Taylor, 2003). In Ho Chi Minh City (HCMC), however, cars are still relative newcomers, at least in their current numbers, and it is only in the last few years or so that these *reconfigurers* of urban space have begun to unwind and re-stitch the existing urban fabric, changing relations, and introducing new timespaces and practices into the traffic system.

Recently, whilst taking my daily walk around the neighbourhood, my path was suddenly blocked by a large SUV car that was involved in the kinds of complex manoeuvres that one might expect of a large object that can only travel forwards or backwards, attempting to turn back the way it had come within a small and cluttered space. The footpath next to me functioned as a multipurpose motorcycle parking area, café seating area, food vendor popup space, and motorcycle repair shop, but this left no space for pedestrians. Therefore, I, as a pedestrian 'going for a walk,' was forced to walk along the road, a practice by no means uncommon in HCMC, and often in crowded streets such as this one, often narrow, and in this case, lined with many cheap fruit and vegetable markets, restaurants, hair salons and cafés, crowded with shoppers, diners, and the hustling and bustling of traffic.

These kinds of traffic conditions in Vietnam have even prompted the following warning on the U.S. Overseas Security Advisory Council (OSAC) website (2017), which warns, not of murder or terrorism, as might be the case with some countries, but of pedestrian sidewalks:

> The combination of a chaotic road system and complete disregard for traffic laws make crossing the street and driving/riding in traffic two of the most dangerous activities in Vietnam. Police are unable to control the rapidly increasing numbers of vehicles on the road . . . The number

of traffic enforcement police is insufficient to deal with the number of vehicles on the road . . . poorly maintained sidewalks, inadequate traffic controls, and the common practice of using sidewalks as a speed lane or a parking space for motor scooters creates a precarious environment for pedestrians. (para. 17)

Denied any passage on the footpath and, given that the large SUV car in front was blocking all traffic with its turning manoeuvres, I ventured further out onto the road when I heard a voice shouting angrily from behind me, "How can you do this? This is car!" I turned to see it was the driver of a car coming from behind, who had slowed down enough to shout at me from inside the car as she passed by. In order to add some context to such an event, it should be noted that HCMC streets are famously chaotic and crowded. The middle of a HCMC street, especially a small one such as this, is a shared space, where sleeping dogs lie, romp and play, humans romp and play, football and badminton games are won and lost, motorcycles and cars stop suddenly without notice, or check their phones whilst driving at a snail's pace, marquees are erected for three-day funerals, joggers jog, pedestrians cross, and motorcyclists shop, all in together; in short, this is not the German Autobahn we are describing here. The modus operandi for all is simply to maintain mobility in any way possible, which often means that the traffic mixes up in all kinds of ways, such as travelling the wrong way on the street, driving along the sidewalk, or, as was my case, walking along the road. No one expects for a moment an obstacle-free experience, but everyone reserves the right to find another way through, even if it means being flexible, so the reaction of this driver surprised me. I imagined myself riding

Figure 7.1 Screenshot from video footage taken by Ngân, walking the HCMC streets while working as a lottery ticket seller, 2016.

my motorbike and shouting at a pedestrian in front of me, "How can you?! This is motorbike!" only to realize that it just didn't work in quite the same way.

What I suspect occurred was that the driver had committed Whitehead's *fallacy of misplaced concreteness*, which, we should recall, is defined as, "the expression of more concrete facts under the guise of very abstract logical constructions" (Whitehead, 1948, p. 52). Given the idealist status of cars (and, therefore, by implication, also of car drivers), in Vietnam right now, it is just possible that this driver had displaced the actual concreteness of the HCMC streets for a more ideal – and, perhaps, more autobahn-like – version of a road that would better suit the equally idealized and abstract version of the car she was driving. If this was so, it still begs the question as to why this driver, who, up until then, had successfully manoeuvred around dogs, funeral marquees, flying shuttlecocks, potholes, low hanging cables, various kinds of concrete and steel protuberances, and cars, bikes, planes, trains, and automobiles, chose this moment to express her surprise at the obstacle that was the present author (other than because I am simply the slowest thing on the roads and, therefore, present as an easy target).

A car, like any other object, has multiple realities and exists in multiple realms. Though it exists as an actual concrete thing, emerging from out of complex events that involve interdependent relations with other things like shuttlecocks and street dogs, our access to this complex relational event/thing/process comes to us, via processes of limitation, as an abstraction, as 'car.' A car exists, on one hand, as a non-perceivable ideal, known only through logic, residing in the eternal realm beyond sensory perception, as a Platonic form, like the idea of a perfect circle, and, on the other, as a perceivable, imperfect, concrete thing emerging from out of events in the actual world (Stebbing, 1926). Did the shouting driver mistake her actual imperfect concrete car (as well as the imperfect road to drive it on) for an eternal geometrical form, a perfect circle, an ideal car? A car, as an event, is constituted of relations that knit together the virtual and the actual, and as Slater states, a car "is not a car because of its physicality but because systems of provision and categories of things are 'materialised' in a stable form" (Slater, 2002, p. 101).

In Vietnam right now, there appears to be a growing aura of something akin to a Platonic idealism around the notion of 'car' that can, to some extent, displace the real and the concrete, with abstractions constituted via other, more idealist relations, upon which further abstractions, such as 'smart', 'modern', 'educated', 'future', and 'urban' then become entangled. As Žižek (2017) notes, universal essentialism of this kind does have something of an *in-itself* existence, though not static, but in a dynamic, changing form. Žižek uses the example of 'communism,' an abstract and ideal entity, that gets constituted through relations and actualized through practices, but continues to have an excess or being beyond concrete actuality.

Recent public statements in Vietnamese media appear to vilify the hard-working and loyal motorbike, the vehicle that Vietnam was built on the back of, in favour of cars, which are viewed increasingly as an essential ingredient in the emergence of a modern 'smart' city. An article published in Vietnamese language on the traffic website, *Roads*, the official site of the directorate for Roads of Vietnam, recently said that motorcycles are one of the anchors slowing down the development of Vietnam due to the ways they "make" or "influence" people to think and behave like petty "peasants," creating negative consequences for the economy and for society in general (Nguyen, 2015, para. 2).

This is quite a lot for the unassuming motorbike to take on board. However, the issue is not so much concerned with whether or not these so-called 'peasant,' 'rural,' or 'unmodern' practices exist, but whether these practices inhibit economic growth and 'progress,' whether they are, in fact, not 'modern,' and whether it is the motorbike-practices-infrastructure that creates and maintains these practices, or whether they would still endure within car-practices-infrastructure. Motorcycles in Vietnam have always afforded certain movements that allow flexibility in the traffic system, and infrastructures have grown up around these affordances, such as the convenience of stopping to shop anywhere or the ability to suddenly change direction or stop without the need to consider how this might affect others in the nearby physical vicinity. However, these kinds of practices continue to persist in non-motorbike vehicles, because car drivers and also truck, bus, and van drivers still have the same kinds of expectations around freedom and fluidity, coupled with a persisting conceptual/physical sense of spatiality that has endured from the days when they rode motorcycles. Cars are, statistically, still very much in the minority in HCMC, but their fast-growing numbers now dominate the traffic as an increasingly brute presence, and the effect of so many cars appearing on the roads in recent years has been nothing short of spectacular in how it has reconfigured relations in the traffic.

The suggestion that motorcycles are an 'anchor' for Vietnam's progress runs contrary to concrete actuality, as it is mostly due to the fluidity afforded by motorcycles, which are still, by far, the majority, that the HCMC traffic persists in being mobile at all. Interestingly, it was not so long ago that motorcycles, rather than cars, were imbued with a similar aura of the 'modern' and the 'cosmopolitan.' One Vietnamese writer recalls a time when there were much fewer motorcycles in Vietnam and those that owned one, especially in the provinces, often had a family member studying overseas. According to Phạm (2012), this meant that the intellectual level of these people was higher – not just the person studying overseas, but the whole family, and so the unassuming motorcycle was then viewed as a symbol of internationalism and of higher levels of academic success.

This idealist conception of the 'car' in Vietnam is another example of Whitehead's fallacy of misplaced concreteness, as the actual becomes displaced by abstractions that are increasingly divorced from concrete reality. The danger in falling prey to this fallacy is that the logic of the ideal, rather than the logic of the real, comes to act like an attractor, around which other things gather, influencing policy, real-world decisions, and expectations. It is slightly ironic to view motorcycles as an anchor holding back the economic and social development of Vietnam when it is motorbikes that allow the transportation of smaller loads at a much cheaper cost, and the existence of thousands of motorcycle couriers in HCMC, ferrying all kinds of products, from fridges, to people, to takeaway food and individual cups of coffee, is evidence of this need. The fluidity created by the large numbers of these more manoeuvrable vehicles and their propensity for slipping through small spaces is the fundamental reason why the traffic in HCMC is not at a standstill, as is the case in some other South-East Asian cities such as Bangkok and Jakarta.

In 2010, a Vietnamese government poster appeared encouraging people to participate in the economic development of Vietnam. In the bottom right corner of the poster, there are three people with their right hands held high in a kind of salute, facing towards the top left corner, where a Vietnamese flag and the Communism flag are featured. The three figures depict a young girl holding a book and wearing traditional clothing (a student), a man wearing only a pair of shorts and holding a large fish (a fisherman?), and an older woman holding some kind of vegetable, a farmer or rural worker of some kind. In the background, there is a large bridge spanning an expanse of water, the colour of which is the blue of whales and oceans in children's fairy tale books. The centrepiece of this poster, however, is the traffic on the bridge. The scene depicts brightly coloured cars, evenly and generously spaced apart, but interestingly, there are no motorbikes amongst the group. Though nine years old, the writing (or, should I say, the picture) is already on the wall, and this poster, as a representation of linear progress, makes it clear that the future is cars rather than motorbikes. Currently, it seems as though the whole of HCMC is becoming divided into the 'car-haves' and the 'car-have-nots,' between which, the line is much more entangled than many might imagine, but, nevertheless, is manifesting, not only in the ideological realm, but also as vastly different qualities of mobilities experiences.

The 'wicked' problems of increasingly crowded urban traffic environments and worsening levels of air pollution present major challenges for policy-makers in Vietnam, of which there are no simple solutions. On March 19, 2019, the Director of the Department of Transport in Hanoi publicly announced plans to prohibit both motorcycles and cars from entering many districts in both Hanoi and HCMC (Trường Phong, 2019). The banning of motorcycles is problematic, as the vast majority of HCMC residents rely on them for mobility, especially given the current public transportation options, and the motorbike is deeply embedded in Vietnamese ways of life. In response to concerns from the public about the choice of options available to traffic users if motorcycles were banned, the transport minister suggested that those who have money could take a taxi and those who do not could ride a bicycle.

The proposal, submitted in early 2019, intends to promote public transport and control private motorized vehicles to inner city districts, including a ban on motorcycles by 2030 and charging fees for automobiles entering the areas by 2025 (Le, 2019). Though the plan also includes the addition of new bus lines, with limited space to build dedicated bus lanes, buses may still be faced with streets gridlocked with cars, for whom a toll may not represent enough of a deterrent. This new proposal aligns with, and is intended to support, the sustainable development goals and targets, as outlined in the *Vietnam National Action Plan for the Implementation of the 2030 Sustainable Development Agenda* (Office of the Government, 2017). Of course, at this stage, this is only a proposal, and it is, no doubt, a smart move for the Vietnamese government to put in writing such intentions, as it would also satisfy certain criteria of international organizations such as the United Nations. The proposal may never be actualized, or may be actualized in a different form, but, nevertheless, publicized intentions such as these can inject anxiety into the general populace, who can never be sure whether or not they will become concrete fact. In fact, the HCMC People's Committee Vice Chairman Tran Vinh Tuyen said the city would "not ban motorbikes because they're a means for people to commute and to work" (Hoang Thuy, 2019, para. 2). Though the proposal does suggest restricting and eventually banning motorbikes from several downtown areas from 2025, according to Tuyen, it is more about encouraging people to choose public transport options, and only once these options offer sufficient convenience, rather than an outright ban on motorbikes (Hoang Thuy, 2019). However, Tuyen also said, "there is no country where traffic is smooth with lots of motorbikes" (Hoang Thuy, 2019, para. 7), and that this is why it is necessary to restrict them, though I wonder if Vietnam, with its unique history of motorcycles, could also lead the way in showing the world how a modern 'smart' city might integrate millions of motorcycles into a smooth urban traffic system.

Edensor notes that the rhythms inside of the car, which may lull passengers and drivers into a state of kinaesthetic relaxation, are different from the rhythmic environment external to the car, and, in the case of HCMC traffic, this contrast is greatly magnified (Edensor, 2010). Driving a car in HCMC is like sitting in a glass and metal bubble, while moving through a swarm of bees, an environment filled with things darting this way and that, constantly impeding your path. Whilst watching the traffic video in a video viewing session, taken by a Vietnamese driver, Thanh – a long-term resident of HCMC, but a relatively new car driver – we watched as motorcycles, one after the other, turned in front of his car, even though he had the legal right to turn before them. He commented on how motorcycles just go whenever there is space and do not care whether they should give way to others and said that this perceived right is deeply rooted in their minds that cars should give way to motorcycles. However, it should also be noted that this data was collected two years before this book was written, when cars seemed much more cautious, but as their numbers have increased, other emergent outcomes have formed through the mass consensus of more and more cars on the roads, resulting in changes in practices and expectations.

Problems in ambient conditions such as air pollution and noise pollution are increasingly impacting the quality of peoples' lives in HCMC, and those who travel in mobilized air-conditioned cabins (cars) are able to avoid the high concentrations of harmful pollutants in the air, created by the large number of motorized vehicles, amongst other causes (Phung et al., 2016). Current air pollution levels in HCMC are resulting in health problems for millions of people, and more than 90% of children under the age of five years are said to suffer from different respiratory illnesses (Ho, Clappier, & Golay, 2011). Motorcyclists, as well as being many times more susceptible to serious traffic accidents, are exposed to these high levels of toxic pollution, as well as dust, noise, and the effects of the sun and rain, from which the kinds of face masks most commonly used offer little or no protection.

Adding to this situation, cars are simply not designed for the streets of HCMC, where most of the streets are dense networks of narrow alleyways known as *hẻm* (see Figure 7.2), within which, about 85% of the city's inhabitants live (Gibert & Pham, 2016).

These chaotic and seemingly random structures grew spontaneously and organically through a natural process of densification, and therefore the layout is very different from the grid laid down by colonial powers in some areas such as in Districts 1 and 3 (Gibert & Pham, 2016). Also, the design of cars in no way takes into account the fragility of the human leg in the ever-decreasing margins for error that occur between motorcyclists and fast-moving cars in HCMC. Modern cars are designed to drive fast, and

Figure 7.2 The small alleyways known as *hẻm* are sometimes not much more than the width of a motorbike. Screenshot from video footage from Pham, working as a *xe ôm* rider, 2016.

'fast,' in most situations in HCMC, might be only about 30 kilometres per hour. Also, motorcyclists are often forced into dangerous situations, such as having to ride through the narrow available spaces left, as the increasing numbers of cars begin to lock them out of the system, or when larger vehicles diagonally cross through motorcycle lanes. On April 25, 2019, a funeral was held in Hanoi, Vietnam's second biggest city, to remember a street sweeper who was hit and killed by a car three days earlier. In a show of solidarity, the funeral was attended by many other street cleaners and Grab taxi motorcyclists, all proudly wearing the uniform of their occupations in order to highlight the growing menace caused by cars for such workers (Tuoi Tre News, 2019).

Ngân, who is a lottery ticket seller in HCMC, expressed to me her fear in having to walk the streets every day, due to the inadequate infrastructure, such as uneven or inaccessible sidewalks and the chaotic movements of vehicles. She said she is constantly forced to walk in amongst the flow of traffic as the sidewalks are very often blocked with parked motorcycles and vendors and she lives in constant fear of being hit by vehicles. Ngân walks the streets for 10–12 hours every day, and the various challenges of this pedestrian life are made even more difficult for her as she also suffers from leg problems. Wearing a GoPro camera on a chest harness, Ngân collected video footage that reveals the many different kinds of streets she walks through, often deep into the chaotic, but often-times, comparatively peaceful, *hẻm* structures. At some points, the video showed her walking against the traffic lights on large busy roads, as motorcycles sped and weaved around her, missing her by only centimetres (see Figure 7.4). She explained that she lives in constant fear of being hit by vehicles and of falling down, as the streets are "rough"

Figure 7.3 Street cleaners. Screenshot from video footage taken from Phúc's delivery truck, 2016.

Figure 7.4 Crossing the road against the traffic amid the wavy movements of motorbikes. Screenshot from video footage taken by Ngân, walking the HCMC streets while working as a lottery ticket seller, 2016.

with "stuff all over the place" and with "vehicles parking." Yet, despite this, Ngân can be heard often singing in her video, a fact that she remained very shy about when asked. I also asked her whether she prefers the large streets or small streets, and she replied that in large streets, she is "scared of drivers with wavy movements" and on small roads she is scared of being robbed. Ngân, speaking in Vietnamese, constantly used terms translated as 'scary' and 'scared' in her narration of the traffic video and described the traffic as a tiger, because, she said, "it wants to hit me."

Schrödinger's car

Whilst all traffic systems are always in transitional moments, the traffic in HCMC is currently experiencing transitional change that would be described as paradigmatic. For this reason, transitional objects blur themselves across categories, not so easily categorized as a definite 'this' or a 'that.' Like the cat in Schrödinger's famous thought experiment, whose qualities remain "smeared" (Schrödinger & Trimmer, 1980, p. 328) across all states from aliveness to deadness until the box containing the cat is opened to prove one state over another, cars in HCMC exist in multiple modes or states simultaneously. These mobile amalgamations integrate the competences, meanings, and subjective forms of bicycles, motorcycles, and even boats into one fluid stability, all enfolded into the *blackboxed* abstraction we call a 'car.' In other words, whilst the cars in HCMC visually resemble the 'car' on the streets of many other cities, they are often driven in ways that retain the practices associated with previously dominant forms of transport in both rural and urban Vietnam (as do motorcycles, which often ride two or three abreast so that the riders can chat together, even on major highways, a lingering practice retained, I suspect, from the days of bicycles).

Latour (1999) describes "blackboxing" as a process in which all of the complex work and connections that go into the making of a complex object, involving both actors and artefacts, becomes opaque (p. 183). *Blackboxing* creates settled matters of fact (Latour, 1999), a kind of technical delegation, an event or a moment in which all frames of reference in time and space have been collapsed or folded into a single point of effect, presenting as seamless entities with hidden internal complexities. When a *blackboxed* abstraction appears as seamless, we may remain unaware of the entanglements of absent actors such as the engineers, designers, factories and factory workers, marketers, advertisers, supply chains, organizations, and companies, all, in some way, represented or *inscribed* into the object and all part of the network through which the object emerges as a nexus effect. In this way, cars in HCMC may present as simple, independent objects sitting in space, but this view belies all of the complex and on-going work that goes into their being what they are, as points and effects of enfolded pasts and futures.

Some Vietnamese scholars have suggested strong links between river transport and contemporary traffic practices in HCMC. In the journal paper *Đặc điểm giao thông Việt Nam từ góc nhìn văn hóa* ("Traffic Features in Vietnam from the Cultural Perspective"), Phạm (2012) points out that Vietnam is filled with waterways and that the main mode of transport was by boat for thousands of years. According to Phạm, Vietnamese people then developed the practice of following the waterway or the wind direction and are historically more familiar with situations free of restrictive physical lanes that might inhibit movement. Phạm also makes the point that historically, the evolution of the traffic in Vietnam proceeds from origins different to the traffic in Euro-American countries, where traffic evolved from the horse and carriage as the dominant form of transportation.

Such stories of path-dependence, in regard to the adoption of new technologies, are not unusual, especially in the Sociology of Science and Technology Studies area, and the notion of path-dependence has particular relevance for understanding the evolution of traffic systems (Shove, Pantzar, & Watson, 2012). The term "path-dependence" has been described as a phenomenon, whereby small historical events are not "averaged away" (Arthur, 1989, p. 117) by the system, but become fundamental in deciding future outcomes, and whereby certain practices and technologies become progressively 'locked in,' sometimes resulting in inflexibility or inefficiency in the system. Therefore, the notion that contemporary traffic practices in HCMC are still fundamentally influenced by river-faring practices *does*, as surprising as it might seem, hold some water.

Shove notes that many elements constituting traffic in contemporary systems in Western countries dominated by cars were in existence antecedent to the car and that there exist clear continuities, though less evident today, that lead back to horse riding, cycling, machine operation, and also sea faring (Shove et al., 2012). Shove suggests that the horse-to-car trajectory is so influential, especially, in regard to materials and competences, that "the only really new element of car-driving in the USA in the early 1900s was the gasoline engine itself, along with knowledge of how to maintain and repair it" (Shove et al., 2012, p. 27).

Given the scarcity of space and the perception of scarcity of time in the HCMC traffic, humans are not only often viewed as mere objects in traffic by other traffic users, but are often viewed as objects getting in the way of the aims and goals of other traffic users. Objects are relational processes, and the ways in which humans become integrated into these processes, how we encounter and dwell within objects, should be a central consideration in urban traffic research. In *Being and Time*, Heidegger (1996) distinguishes between three different modes of being that describe the ways in which we have experiences with objects: the *present-at-hand* (*vorhandaheit*), the *ready-to-hand* (*zuhandeheit*), and *Dasein*. Objects are encountered in the mode of *present-at-hand* when we engage with them in a theoretical or analytical way, such as when we focus on their properties in an objective or "scientific" way. Such an encounter with an object usually occurs under 'special' circumstances, such as in cases of a broken tool, whereby we need to examine the tool more closely, in which case, the entity, which would normally recede into the background behind a cloak of instrumentalism, 'pops' out in an existential spatial *nearness* for us.

However, probably the most common way we encounter objects is in the mode of *ready-to-hand*, whereby we are just using the tool in a practical way, absorbed in the action and focused on the goals that reside in the future, the kind of immersion in praxis that Dreyfus calls *absorbed coping* (Dreyfus, 2007). When the tool is working seamlessly and we are in the flow, the tool tends to recede into the background of what Heidegger calls the realm of *tool-being*. From the perspective of the driver who shouted at me in the earlier story, I probably appeared as an object encountered in the mode of *present-at-hand*. Objects in the mode of the present-at-hand are encountered as "unready-to-hand" or "unhandy" (Heidegger, 1996). According to Heidegger (1996), this can occur in the following modes:

- Conspicuousness: when tools are damaged or unsuited to the job;
- Obtrusiveness: when tools are missing; or
- Obstinacy: when something gets in the way of us taking care of things.

When tools are working seamlessly for us, they 'disappear' into the background, and we work *through* them, on our way to something or somewhere else, towards our goals, as when the windscreen wipers directly in front of our face are no longer really 'present' for us, though they may be mere centimetres from our nose. According to Polanyi (2009), this is the structure of tacit knowledge: we attend *from* something in order to attend *to* something, and the something we are attending *from* becomes almost invisible due to our focus on this further goal or destination. Due to the crowded and increasingly busy character of the traffic in HCMC, objects get noticed when they emerge from the background as *present-at-hand*, popping out as an obstinate obstruction, and this is an increasingly common phenomenon.

Heidegger (2001) views objects as gatherings of relations that emerge through relations and processes of the existential being of people engaged in absorbed coping, immersed in trajectories towards aims and goals. Following Heidegger's (2001) notion of the "gathering" (p. 171), actor-network theory (ANT), as described by Law (2002), also views objects as emerging as constituted by relations, as an "effect of stable arrays or networks of relations" (p. 91), as "relational contingencies" (p. 92). ANT also imagines objects as multiplicities, existing in more than one kind of spatiality simultaneously, as Euclidean, Cartesian, and topographical spatial entities in network spaces. As Law and Singleton (2005) explain, when the same object (they use medical examples of 'objects' such as 'alcoholic liver disease' and anaemia, all objects that change form and nature in different contexts) is constituted with different relations, it is "no longer a matter of different perspectives on a single object but the enactment of *different objects* in the different sets of relations and contexts of practice" (p. 342). Similarly, Mol (2002) describes how the disease, 'lower-limb atherosclerosis,' exists as one entity in the form of the textbook object, yet also emerges as a quite different object in other actualizations. Whereas ANT (at least in its earlier incarnations) views the world as stable immutable network objects, Whitehead's metaphysics, which posits a stratified nature of constant birthing and perishing of entities, presents as more fluid and dynamic, focusing on instability, rather than stability, on the modes rather than the nodes. Noting this aspect of ANT, Thrift (2008) has said that

> Actor-network theory is much more able to describe steely accumulation than lightning strikes, sustained longings and strategies rather than the sharp movements that may also pierce our dreams. (p. 110)

However, it should also be noted that more recent incarnations of ANT as 'post-ANT' have, as Law and Singleton note (2005), "loosened up on networks, considered fluidities and explored the ambivalences" (p. 341).

As a tool for understanding how scientific objects hold their shape, as well as for thinking about long-distance power structures, Latour (1986) formulated the idea of the *immutable mobile*, an object that retains its form in one spatiality (immutable and fixed in a network space), whilst being mobile in another, for example, in a physical space. It is for this reason, that spatiality, objects, and notions of mobility cannot be considered independently from each other. For example, a Boeing 747 is able to fly (Latour, 1999), and a ship is able to sail (Law, 1987) in topographical or physical space, contingent only on its stable, *immutable* constitution in network space. In Latour's view, a ship and a plane, and by implication, other modern complex technological objects such as a motorcycle, a car, and a scientific fact are all examples of what Latour (1986) calls "immutable mobiles" (p. 7), because their fixed configurations or *immutability* in network space is the very thing that allows them to be mobile in another kind of spatiality (Latour, 1986; Law, 1999, 2002; Law & Singleton, 2005). A car, for example, is mobile in a physical space dependent upon its relational networked constitution within structures that include traffic lights, gas stations, tyre factories, traffic laws, and the offices of traffic engineers, etc. However, to what degree is the object really immutable? What kind of fixedness is required for such immutability, and in what way? In the nodes or outcomes on the ends of relations, the order and structure of the relations, or the nature of the relations themselves, all of which are interdependent?

An example of an immutable mobile in HCMC traffic might be the bicyclists who ride around the city collecting refuse, such as recyclable scrap metal, cardboard, etc. HCMC has, arguably, one of the most comprehensive and efficient, as well as the most informal, recycling programs of any similarly large city. Women (they are always women) on bicycles or pushing carts on the roads collect and transport large loads of scrap refuse of all kinds. This job, in such a dangerous and busy traffic environment, requires enormous persistence, patience, and courage, and it appears that age is no barrier. In one sense, these bicyclists are mobile, in that they are moving physically along the road, but in another, they are immobile, fixed, to a degree, as a nexus of relations that traps them in a cycle of replicated actions and reactions. In other words, the very relations that allow these women to be physically mobile also keep them from improving the quality of their lives, and there are power structures and network configurations in place that have a vested interest in keeping these networks stable.

If a car has the character of *carness*, a motorcycle, *motorcycleness*, or a bicycle, *bicycleness*, where might this emergent character reside and how might it endure within fluid relational processes? Law and Mol (2001) describe the existence of what they call "fluid" objects and spatialities of moving and shifting borders that retain a core identity, whilst being reconstituted through different relations, different practices, and in different physical locations. One such object is the Zimbabwe Bush Pump, a fluid technology that is "solid and mechanical" (p. 225), but with vague, moving boundaries, whilst maintaining a core identity of *Zimbabwe-Bush-Pumpness*. The Zimbabwe Bush Pump, as described by de Laet and Mol (2000), is a device that maintains continuity in its homeomorphic self, even through different configurations. The device is able to change incrementally as bits and pieces are replaced in ad hoc, piecemeal ways, and even the definition of what it means to be doing its job changes from performance to performance and place to place. Nevertheless, there is continuity in its essence, even while this essence is fluid and changing. Of their experiences with this object, de Laet and Mol (2000) write:

> We find that in travelling to intractable places, an object that isn't too rigorously bounded, that doesn't impose itself but tries to serve, that is adaptable, flexible and responsive - in short, a fluid object - may well prove to be stronger than one which is firm. (p. 225)

Examples of these kinds of objects are also commonly found on the streets of HCMC due to the pragmatic nature of people often just having to make do with what they have. One example of these are the mobile "wet" markets transported on motorcycles (see Figures 7.5 and 7.6) that retain an essential identity even whilst being constituted and reconstituted, unpacked and packed, in different ways. These mobile markets, often transporting live animals such as fish, which are then killed and dissected on the street, along with all kinds of meat and vegetables, also cut, weighed, and packaged for customers on the spot, form pop-up communities, acting as centres for the purveyance of, not only food, but also of news and gossip. Watching the setting up of these mobile markets from my apartment window is like peeking inside a blackbox, as the relations become more visible, unpacked, and disentangled.

Viewing objects as constituted across different spatialities is a productive tool in seeing how encounters with objects come to be imbued with meaning and how objects, space, and time emerge interdependently, embedded in infrastructures of meaningful relations. Laura, originally from Eastern Europe, rides a motorcycle and has lived in HCMC for about two months, and Duy is a government bus driver with many years of experience in the HCMC traffic. Together, these two traffic users viewed a traffic video that

Figure 7.5 Mobile motorcycle market. Copyright 2017 by G. Wyatt.

Figure 7.6 Mobile motorcycle market set up on the street. Copyright 2017 by G. Wyatt.

happened to feature a two-wheeled cart attached to a motorcycle, known in Vietnamese as a *xe ba gác* (see Figures 7.7 and 7.8). Duy pointed out that this vehicle is unique to Vietnam and has been "redesigned" here, but Duy was also wary of these vehicles and commented:

> From what I understand, they are redesigned, so there is no guarantee that they are safe to operate with other vehicles, in terms of technical factors. A wheel may fall down or the frame may be broken at any time.

For Duy, this vehicle is encountered as a kind of matters-of-concern gathering, a potentially dangerous nexus of relations that includes insufficient regulations around traffic safety and other historical factors. He said, "other traffic users when driving near them [near the *xe ba gác*] would feel very scared as they may cause some danger." However, for Laura, who is new to HCMC, this *xe ba gác* was merely encountered as an interesting Vietnamese artefact, rather than this affective danger-in-potential experience of Duy's. For the bus driver, the encounter is informed by virtual movement, as virtual lived relations, described by Massumi (2008) as "action appearing in potential" (p. 5). The vehicle, being an event, is filled with what Massumi (2010) calls "affective facts" (p. 54), such as the affective potentiality of a wheel falling off at any moment or the frame breaking. These attributes, qualities, or properties of the vehicle, though they reside in an affective future, as 'action appearing in potential,' nevertheless concretely influence the present and therefore should be seen as constituent of the object as its steely materiality.

Figure 7.7 Xe ba gác. Copyright 2019 by G. Wyatt.

Figure 7.8 Xe ba gác. Copyright 2019 by G. Wyatt.

The fluid way the *xe ba gác* has been put together, with a focus on pragmatism and interchangeable but unified bits and pieces (a fluid object), reflects some of the same kinds of processes and structuralities found both in and beyond the domain of traffic. It is, therefore, not the simple object that it appears to be, and is constituted within entanglements of historical factors, future projections, economics, pragmatism, technology, materiality, creativity, politics, and environmental influences. The *xe ba gác*, then, is like a localized resonance, an intersection of almost infinite other occasions, formed in its own unique way as an event from its own standpoint. The *xe ba gác*, also, mirrors that kind of stubborn but fluid resilience characteristic in many aspects of Vietnam. The ability of this vehicle to remain an enduring form on the streets of HCMC, its mobility, relies on its immutability, but also on its *mutability*. A professional van driver named Viet later told me that there are clear regulations prohibiting these vehicles carrying loads in the city, but these rules are often ignored by drivers, who, weighing up the need to make money against the risks of breaking the law, risk it anyway. Also, various physical components of the *xe ba gác* are easily found and replaced, so that parts may easily be modified and interchanged with others, making it a mutable, fluid object, both in patterns of practices and materiality.

We can now imagine this entity the 'car' in a multitude of forms and enactments: as an ideal eternal form, as a more ephemeral emergence of character (carness), and as a series of events, constituting real (and messy) concrete actuality. What remains now is to dissolve the human and the 'subject' into these event/amalgamations and to close the ontological gaps between subject and object.

References

Arthur, W. B. (1989). Competing technologies, increasing returns, and lock-in by historical events. *The Economic Journal, 99*(394), 116–131.

Dant, T., & Martin, P. (2001). By car: Carrying modern society. In J. Gronow & A. Warde (Eds.), *Ordinary consumption* (pp. 143–57). London, United Kingdom: Routledge.

De Laet, M., & Mol, A. (2000). The Zimbabwe bush pump: Mechanics of a fluid technology. *Social Studies of Science, 30*(2), 225–263.

Dreyfus, H. L. (2007). Philosophy 185 Heidegger (Audio podcast). Retrieved from https://archive.org/details/Philosophy_185_Fall_2007_UC_Berkeley

Edensor, T. (2010). Introduction: Thinking about rhythm and space. In T. Edensor (Ed.), *Geographies of rhythm: Nature, place, mobilities and bodies* (pp. 1–18). Farnham, United Kingdom: Ashgate Publishing, Ltd.

Gibert, M., & Pham, T. S. (2016). Understanding the Vietnamese urban fabric from the inside. *The Newsletter, 73*, 32–33.

Heidegger, M. (1996). *Being and time: A translation of sein und zeit* (J. Stambaugh, Trans.). New York: State University of New York Press.

Heidegger, M. (2001). *Poetry, language, thought* (A. Hofstadter, Trans.). New York, NY: Harper Perennial Classics.

Ho, B. Q., Clappier, A., & Golay, F. (2011). Air pollution forecast for Ho Chi Minh City, Vietnam in 2015 and 2020. *Air Quality, Atmosphere & Health, 4*(2), 145–158.

Hoang Thuy. (2019, March 7). HCMC will not ban motorbikes, promises leader. *VNExpress*. Retrieved from https://e.vnexpress.net/news/news/hcmc-will-not-ban-motorbikes-promises-leader-3891039.html

Latour, B. (1986). Visualization and cognition. In E. Long & H. Kuklick (Eds.), *Knowledge and society: Studies in the sociology of culture past and present* (Vol. 6, pp. 1–40). Retrieved from http://www.bruno-latour.fr/sites/default/files/21-DRAWING-THINGS-TOGETHER-GB.pdf

Latour, B. (1999). *Pandora's hope: Essays on the reality of science studies.* Cambridge, MA: Harvard University Press.

Law, J. (1987). Technology and heterogeneous engineering: The case of Portuguese expansion. In W. E. Bijker, T. P. Hughes & T. J. Pinch (Eds.), *The social construction of technological systems: New directions in the sociology and history of technology* (pp. 111–134). Cambridge, MA: MIT Press.

Law, J. (1999). After ANT: Complexity, naming and topology. *The Sociological Review, 47* (1_suppl), 1–14.

Law, J. (2002). Objects and spaces. *Theory, Culture & Society, 19*(5–6), 91–105.

Law, J., & Mol, A. (2001). Situating technoscience: An inquiry into spatialities. *Environment and Planning D: Society and Space, 19*(5), 609–621.

Law, J., & Singleton, V. (2005). Object lessons. *Organization, 12*(3), 331–355.

Le, V. H. (2019, March 2). HCMC inner districts to ban motorbikes by 2030. *Viet Nam News.* Retrieved from https://vietnamnews.vn/society/505700/hcmc-inner-districts-to-ban-motorbikes-by-2030.html#7VMMBu0ujpbR9WSr.97

Massumi, B. (2008). The thinking-feeling of what happens. *Inflexions*, 1, 1–40. Retrieved from http://www.inflexions.org/n1_The-Thinking-Feeling-of-What-Happens-by-Brian-Massumi.pdf

Massumi, B. (2010). The future birth of the affective fact: The political ontology of threat. In M. Gregg & G. J. Seigworth (Eds.), *The affect theory reader* (pp. 52–70). Durham, NC: Duke University Press.

Mol, A. (2002). *The body multiple: Ontology in medical practice.* Durham, NC: Duke University Press.

Nguyen, D. T. (2015, March 11). Tư duy tiểu nông và nền văn hóa xe máy [The mindset of peasants and bike culture]. Retrieved from https://www.linkedin.com/pulse/t%C6%B0-duy-ti%E1%BB%83u-n%C3%B4ng-v%C3%A0-n%E1%BB%81n-v%C4%83n-h%C3%B3a-xe-m%C3%A1y-duc-truong-nguyen/

Office of the Government. (2017). *National action plan for the implementation of the 2030 sustainable development agenda.* Retrieved from http://www.un.org.vn/en/publications/doc_details/543-the-national-action-plan-for-the-implementation-of-the-2030-sustainable-development-agenda.html

Overseas Security Advisory Council. (2017). *Vietnam 2017 crime & safety report: Ho Chi Minh City.* Retrieved from https://www.osac.gov/Pages/ContentReportDetails.aspx?cid=21343

Phạm, N. T. (2012, September 15). Đặc điểm giao thông Việt Nam từ góc nhìn văn hóa [Traffic features in Vietnam from the cultural perspective]. Retrieved from http://vhnt.org.vn/tin-tuc/y-kien-trao-doi/27696/dac-diem-giao-thong-viet-nam-tu-goc-nhin-van-hoa

Phung, D., Hien, T. T., Linh, H. N., Luong, L. M. T., Morawska, L., Chu, C., Binh, N. D., & Thai, P. K. (2016). Air pollution and risk of respiratory and cardiovascular hospitalizations in the most populous city in Vietnam. *Science of the Total Environment, 557–558*, 322–330. Retrieved from https://eprints.qut.edu.au/95857/

Polanyi, M. (2009). *The tacit dimension.* Chicago, IL: University of Chicago Press.

Schrödinger, E., & Trimmer, J. D. (1980). The present situation in quantum mechanics: A translation of Schrödinger's 'cat paradox' paper. *Proceedings of the American Philosophical Society, 124*(5), 323–338.

Shove, E., Pantzar, M., & Watson, M. (2012). *The dynamics of social practice: Everyday life and how it changes.* London, United Kingdom: Sage.

Slater, D. (2002). Markets, materiality and the "new economy". In S. Metcalfe & A. Warde (Eds.), *Market relations and the competitive process* (pp. 95–113). Manchester, United Kingdom: Manchester University Press.

Stebbing, L. S. (1926). Professor Whitehead's "perceptual object". *The Journal of Philosophy, 23*(8), 197–213.

Taylor, N. (2003). The aesthetic experience of traffic in the modern city. *Urban Studies, 40*(8), 1609–1625.

Thrift, N. (2008). *Non-representational theory: Space, politics, affect.* New York, NY: Routledge.

Trường Phong. (2019, March 19). Giám đốc sở Giao thông Hà Nội: 'Ít tiền thì đi xe đạp, nhiều tiền thì đi taxi' [Director of Hanoi Department of Transportation: 'A little money, I ride a bicycle, a lot of money I take a taxi]. *Tiền Phong.* Retrieved from https://www.tienphong.vn/xa-hoi/giam-doc-so-giao-thong-ha-noi-it-tien-thi-di-xe-dap-nhieu-tien-thi-di-taxi-1390779.tpo

Tuoi Tre News. (2019, April 25). Đám tang đẫm nước mắt của chị lao công nghèo [The tearful funeral of the poor worker]. *Tuoi Tre News.* Retrieved from https://tuoitre.vn/dam-tang-dam-nuoc-mat-cua-chi-lao-cong-ngheo-20190425121600505.htm

Whitehead, A. N. (1948). *Science and the modern world.* New York, NY: The New American Library of World Literature.

Žižek, S. & Harman, G. (2017, March 2). *Slavoj Žižek & Graham Harman duel + duet* (Video file). Retrieved from https://www.youtube.com/watch?v=r1PJo_-n2vI

8 The subject in traffic

Lefebvre (2007) does not go quite as far as calling humans 'objects,' but does say that "we *contain* ourselves by concealing the diversity of our rhythms: to ourselves, body and flesh, *we* are **almost objects**" (p. 10). Symmetrical ontologies such as actor-network theory (ANT), object-oriented ontologies, and process philosophies, such as Whitehead's, all aim to circumvent modern philosophy's bias towards the interior world of private sensations and the products of a consciously experiencing subject – namely, a human subject – as the primary data for analysis. Though Heidegger broadened its possibilities, in revealing a world of unconscious experiencing in ontologies of tool-being (Harman, 2002; Heidegger, 1996), phenomenology, with its origins in Husserl and his notion of intentionality, continues to be associated with a human-centred approach. Latour (1999) sums up this attitude in the following:

> Phenomenology deals only with the world-for-a-human-consciousness. It will teach us a lot about how we never distance ourselves from what we see, how we never gaze at a distant spectacle, how we are always immersed in the world's rich and lived texture, but, alas, this knowledge will be of no use in accounting for how things really are, since we will never be able to escape from the narrow focus of human intentionality. Instead of exploring the ways we can shift from standpoint to standpoint, we will always be fixed in the human one. (p. 9)

However, Whitehead's scheme provides valuable opportunities for speculative ways, more empathetic of structures of experience that permeate not only human worlds but also nonhuman and inhuman perspectives. It is, therefore, no surprise that Latour, a founding member of ANT, should draw so substantially from Whitehead.

In an ANT approach towards better understanding processes of agency in human/inhuman collectives (I use the term 'inhuman' here to refer to, for example, technical inscriptions into which a human might be enrolled), all entities are taken on their merits, namely their agentic abilities to bring about change, translate, mediate, and, otherwise, influence the state of things. ANT serves as a tool in unpacking the invisible work that goes on 'behind the scenes' that inform certain power relations, constitute sociotechnical network stabilities, or simply make a thing what it is (Callon, 2001), a relational and ecological ontology that shares similarities with Whitehead.

Object ontologies, such as Harman's, on the other hand, speculate on the inexhaustible lives of 'things,' beyond discourse and beyond their theoretical or practical uses with humans. Harman (2005) is more interested in the independence of the world and the nature of relations between things and objects that lie beyond the scope of any human involvement – the things-in-themselves. When objects recede deeper into the realm of *tool-being*, beyond human experience, beyond their instrumental uses, beyond the epistemological, and irrelevant to the ethical, how then might we speculate as to the nature of these relations? According to Harman (2005, 2007) and Shaviro (2011), the nature of these relationships must then be *aesthetic*, making aesthetics, according to Harman, "first philosophy" (2007). When things cannot be cognitized, conceptualized, utilized, regulated, or defined according to rules, what remains is the open-endedness of aesthetics, style, and character (Shaviro, 2011), and Whitehead's scheme offers a way into an *aesthetic realism* or *ontology of aesthetics*.

Sometimes, a thing can just be surprisingly, simply, (or complexly), and aesthetically beyond cognitive understanding, which involves, as Shaviro (2011) writes, "feeling an object *for its own sake*, beyond those aspects of it that can be understood or used" (p. 7). Harman (2005), following Bergson, regards this as the realm of *sincerity*, linking humour with sincerity. The innate naiveté of the comical, of which all objects have the potential, unearths a world of *aesthetic realism* in which any mundane object can serve as an opening into the 'sincere' aesthetic lives of objects. Harman, like Whitehead, shows how a phenomenological or speculative approach can move beyond the human to encompass a new kind of realism and says that this kind of sincerity "is already the proper meaning of phenomenology's definition of intentionality" (Harman, 2005, p. 135).

The notion of *givenness*, when viewed from the perspective of a more object-oriented ontology, can reveal the kinds of complex manifold connections and relations that constitute our intimate relations with nonhumans in the modern world, as kinds of ontological templates, in that "every form of existence would be explicated in its own language and according to its own condition" (Latour, 2014, p. 305). In other words, *givenness* comes in its own unique languages, modal connections, and *prehensions*, as things are infused with aesthetic character and style. This kind of openness is possible in Whitehead's process philosophy because it retains an "appeal to naïve experience" (Goffey, 2008, p. 15), which is integrated into an entire systematic and analytic phenomenological or speculative approach that bridges realism and idealism.

The ontological openings that lead to the realm of *tool-being* and also to insights about the *sincere* aesthetic nature of objects are to be found everywhere in the mundane. Such points of access, for one equipped with the right sensibilities, bring into view the realm of ambiguity and the ability of objects to continually surprise us, what Harman calls "allure" (2005, p. 141). Shaviro (2011) says that *allure* is an instance of Whitehead's *lures for feeling*, an aesthetic givenness, a subjective aim in the form of an attraction or repulsion towards a thing: enticing, exciting, inciting, seducing, tempting, or otherwise compelling. Harman (2005) cites examples of allure in things that appear underequipped in tasks usually performed with ease, such as a newborn horse that is still becoming familiar with its legs, as contrasted with the expert moves of a galloping adult horse (p. 142). Events such as these reveal surprising sides to objects that circumvent their functional usage or theoretical understanding, but also present a naïve comic sincerity.

The following story illustrates an event in which objects emerge as multiplistic beings through reconfigurations of existing power structures. However, more than this, it gives insight into Whitehead's notion of *lures for feeling*, allowing us to appreciate how mundane objects have the capacity to surprise by *erupting* or *bursting forth* (Shaviro, 2011) in *alluring* ways, whilst, simultaneously retreating, beckoning us to follow into the dark corridors that lay beyond instrumentalism and conscious cognition. In Ho Chi Minh City (HCMC), when it rains, it usually rains heavily and suddenly, like a waterfall, like a large tap suddenly turned on and then, perhaps 30 minutes later, turned off again just as quickly. However, there are those rare cool days when the rain comes on slowly, gradually, *seepingly*, often due to a storm located further north in Vietnam, and then stays most of the day. It was one of these days, with the dark clouds low in the sky and a grey soggy blanket draped over the city, when Nguyên, a HCMC taxi driver, was out driving his taxi. When the rain began falling, Nguyên turned on his windscreen wipers, but soon became aware of the growing presence of an unfamiliar sound.

Car windscreen wipers, like the glasses we wear on our nose, are one of those objects that when working seamlessly, cease to exist, disappearing into the world of tool-being. Though physically close, they may as well be miles from us, hidden amongst the spatiality of the existential, obscured by our aims and goals, which might lay further afield. In contrast to the notion of simple location, Heidegger's (1996) existential spatiality suggests that it is our existential being that informs what exists for us and structures our immersed *being-in-the-world* in terms of nearness and farness. Heidegger (1996) describes an ontological chain of being, using phrases such as "in-order-to," "towards-which," and "for-the-sake-of-which" that provide direction and structure relationalities through our concerns, aims, and goals, therefore imbuing things with relevance and significance that open up fields and horizons, whereby things come into existence and are encountered accordingly. Therefore, windscreen wipers are generally encountered in the mode of the *ready-to-hand*, but should the wipers become *unhandy*, in Heidegger's terms, in the mode of conspicuousness, obtrusiveness, or obstinacy, they will be encountered as such, necessitating an analytical attitude, a situation that occurred in the following story.

Watching Nguyên's traffic video (the GoPro camera was mounted on the dashboard, so the windscreen wipers are in close-up, seemingly swiping across the camera), we can see the wipers begin to drag more slowly across the glass, labouring as the raindrops splash with more intent on the windscreen of the car. Eventually, both wipers stop moving, and we watch as the world outside becomes prismatic, a slow moving distorted landscape. As the cars, trees, and motorcycles beyond the glass blur and morph, Nguyên slowly pulls off the road into a park of trees and cafés.

As he pulls the vehicle to a stop on the side of the road, a female passenger in the back seat can be heard chatting casually with him. About a third of the camera's view is obstructed by the stuck wiper and with the rain now falling heavily, the visual perspective is close and claustrophobic. Suddenly (not unlike one of those 'reality' hand-held camera kind of horror movies), we see Nguyên's hand come into view on one side. He has been forced out into the heavy rain and is trying to manually 'kick-start' the wipers with his hand, which then put in a few half-hearted swipes, only to stop again in the very position they were before. Nguyên moves the car and parks, so that the driver's side is now under a large umbrella situated at an outdoor café and he gets out once more.

The rain is now torrential on the windscreen as it rushes off the side of the café's roof, and the sight of all that water so close makes it difficult to breathe when viewing the video, as though we are underwater. All that water only makes things more difficult, and, in what is fast becoming a splashy comedy of errors, Nguyên moves the car again (for about the fourth time) to a new spot, and, when the rain stops, he leaves the safety of the car and tries again. This time he pulls the left-side wiper up so it stands vertically. As he does this, the wipers begin moving. The left-side wiper is now doing a kind of vertical dance, causing the right-side wiper to speed up and flick dangerously fast across the glass. Nguyên tries to reach across to do the same with the right-side wiper, but the left side, the wiper that is vertical, begins hitting him in the face (see Figure 8.1), so he quickly retreats back to the safety of the car's interior to think things through. After turning off the wipers, he then climbs across the bonnet, pulling up, vertically, the right-side wiper and laying the left side flat. He turns them on again. We now have a situation where the left side is wiping the window (the driver's side is on the left) and the right-side wiper is in a vertical position, swinging around like its Saturday night. And we are off driving again.

The point of this dramatic comic-tragedy of water and wipers is that the reservoirs (excuse the watery puns) of potential that lie within objects can never be exhausted through the limitations of our abstractions and instrumental engagements. Broken tools provide opportunities to view things afresh, as they lure us into previously unknown networks, forcing new kinds

Figure 8.1 Fixing windscreen wipers in the rain. Screenshot from video footage
taken from dashboard camera in taxi, 2016.

of abstractions through new kinds of givenness. It is for this reason that
Harman (2005) suggests, "to revive phenomenology means to restore our
taste for the specific textures and overtones of concrete experience" (p. 21).
This prodigious performance by a dancing wiper infused the whole event
with a certain character and, in doing so, also changed the power relations,
forcing the driver from the safety of the car's interior to the wet wilds of the
exterior. A wiper, when functioning as it was designed to, presents as one
kind of felt relationality or prehension, but a vertically 'dancing' wiper presents as a new kind of prehension.

The notion of a human subject as 'hybrid' or 'cyborg' is a familiar theme
in ANT and in Science and Technology Studies. In the context of traffic, it is
often used to refer to the relationship between, for example, the driver and
the car or the rider and the motorcycle, blending the two entities into a fusion, termed "driver-car" (Dant, 2004). Urry took the notion of the hybrid
in traffic and extended it to encompass the larger assemblage beyond the
vehicle, out into the wider networks of "machines, roads, buildings, signs
and entire cultures of mobility" (Urry, 2006, p. 18). Thrift (2008), following Katz, says to truly imagine the intimacies that occur between a driver
and a car demands a "metaphysical merger" (p. 80), from which emerges
a distinctive ontology, whereby the being of the driver and the being of
the car form an emergent outcome more than the sum of their parts. Latour (1999) skilfully demonstrated such a merger within the context of the
gun issue in the USA, introducing a third agent, a hybrid form, more than
gun, more than shooter, which he called a "citizen-gun" or "gun-citizen"
(p. 179). In this way, Latour uses the notion of the hybrid to explore the

usual ANT agendas of technical mediations and translations of goals, and this example demonstrates the distinctive ontology that arises when the inscriptions of the gun and the aims or agenda of the shooter intertwine and amalgamate.

If we dispense with the independent exclusivity of the human interior world as fundamental to the structuring of an ordered universe, how are we to conceive of the subject and where is the subject located in all of these happenings? In a sense, like the ephemeral ghostly 'object' that atmospherically characterizes events, the subject is, likewise, out in the world, ephemerally distributed, "leaky," to use Manning's (2009) term. Because Whitehead's ontology prioritizes fluidity, form, and change over stasis, the notion of a subject is viewed more as a relational process, an *activity* of the relation, rather than a term of the relation (Venn, 2010). Whitehead retains the term 'subject' because it is familiar in philosophy, but he performs an inversion of the usual Cartesian order of the relationship between consciousness and experience that has dominated philosophy (Goffey, 2008). Instead, Whitehead's 'subject', termed a "subject-superject" emerges from out of the world – "an actual entity is at once the subject experiencing and the superject of its experiences" – (Whitehead, 1978, p. 29) through a process similar to Simondon's *individuation* (Chabot, 2003), in which the subject is neither the starting point nor the end product (Chabot, 2003; Venn, 2010). The term "subject" comes from the Latin, which literally means 'thrown under,' and Whitehead's term 'superject' means "thrown beyond" (Hosinski, 1993, p. 89). In both Simondon's notion of *individuation* and Whitehead's notion of the *superject*, what matters is the nature of the relations and the fact that these relations are not something that occurs once the subject already exists. As Chabot (2003) explains, "the relation does not connect A and B once they have already been constituted. It is operative from the start. It is interior to their being" (p. 77), and no substance can exist without its relations to other entities. To fully comprehend this openness, relations and their subjective forms (forces, energies, lures for feeling) need to be understood in terms of affective engagement and attunement as the common ground of exchange (Venn, 2010).

In traffic, goals, aims, tacit skills and knowledge, embodied awarenesses, *attunements*, and our absorbed coping, are intimately intertwined with non-human and inhuman entities, so that agency and affective engagement is distributed throughout the collective through mediations and translations of goals. According to Harman (2005), the aesthetic presence of objects emerges from their *autonomous* nature as aesthetic things in the world. However, aesthetics, character, and atmosphere – objects that characterize events – are emergent outcomes that result from complex processes that flow far beyond a single appearance, throughout the relations of the collective or assemblage, and so objects, subjects, infrastructures, and environments, like time and space, can never be viewed through reductionist methods or independently.

Bennett (2012) seeks a theory that aims to make such arbitrary borders irrelevant:

> Since everyday, earthly experience routinely identifies some effects as coming from individual objects and some from larger systems . . . why not aim for a theory that toggles between both kinds of magnitudes of "unit"? One would then understand "objects" to be those swirls of matter, energy, and incipience that hold themselves together long enough to vie with the strivings of other objects, including the indeterminate momentum of the throbbing whole. (p. 227)

Subjects and objects, as superjects, must also be seen as emerging from – and always embedded in – the infrastructures that form this throbbing whole. For example, buses are considered extremely problematic by other traffic users in HCMC, exuding a powerful aesthetic and negative influence, like typhoons with wheels, like the running bulls of Pamplona, they part seas of motorcycles in their paths. Ngân, the lottery ticket seller, said, "those big vehicles are really scary. . . . I feel scared of buses because of the careless driving."

Duy, who is a full time government bus driver, collected video footage taken from the front of his bus that showed his entire bus route from the depot and back. Duy is young, and when I met him in the video-viewing session, he seemed friendly, patient, and calm. However, as I watched him watching the traffic video, both his own and the video of another participant (a non-Vietnamese motorcyclist, quite new to the city), I sometimes saw him physically react with noticeable distress to noisy or potentially dangerous scenes, as though he carried the stress of the traffic just below the skin. Whilst most HCMC residents view buses in a negative light, most do not know that bus drivers are given a set time in which to finish their route, after which, if not completed, they are then driving for free; as Lefebvre (2007) says, "power knows how to utilise and manipulate time, dates, time-tables," because it combines the "unfurlings" of individuals, groups, and societies and "rhythms them" (pp. 68–69). HCMC government bus drivers drive 10 hours a day, not including the time spent waiting in the bus between routes. Though this seems extreme, such hours are not uncommon amongst professional drivers in HCMC. Nguyên, the taxi driver, told me that one taxi shift is 24 hours long, of which, 14–15 hours of that time is spent engaged in driving (unless he earns enough money to stop), and that he does 15 such shifts per month.

The time restriction allocated to finishing the bus route results in the emergence of particular kinds of practices and concrete outcomes and creates stress and potential danger for both the bus drivers and other traffic users. Duy, the bus driver, said:

> I only have 80 minutes and 10 minutes more given the traffic jam so I have total 90 minutes. In my opinion, it is not enough if I drive slowly

and carefully. So, I have to drive so quick. Whenever I see a free way, I speed so that I can compensate for the time being in the traffic jam.

As a *subject-superject*, Duy's character, when driving, is inseparable from the relations in which he is embedded, a situation that he appeared to resent. One scene in the video showed a girl on a bicycle in front of the bus on a large road. There are no bicycle lanes in HCMC, so the girl must compete for space with the much larger bus. As we watched the video, I asked Duy to put himself into the mind of the girl in front, to which he replied:

> I think she might think, "There comes an evil." So, they are quite scared. When they hear the bus horn, they may think, "Oh, the evil bus or whatever vehicle is coming" . . . To deal with the traffic in the city, I cannot be a nice person. . . . When there's a traffic jam, I have to quickly get through the area.

Buses use their horns like sonic snowploughs, a practice that contributes substantially to the already immense noise pollution in HCMC. According to Duy, this practice is justified:

> Because the motorcycles – I have to press the horn continuously so that the motorcycles can aware of me. Firstly, they can notice me. Because if I just press the horn once, they are not aware of me, they are not scared. They just keep proceeding. Even if I hit them, they are not scared. So, I have to press in a way that make them feel annoyed a bit so that they can move to the right side to give way for me.

Figure 8.2 Motorbikes often block the paths of buses. Screenshot from video footage taken from government bus, 2016.

Whitehead (1920) says that what we find in space are not substances, but experiential object/attributes, "the red of the rose and the smell of the jasmine and the noise of cannon" (p. 21). In this case, what we find 'in space' is the particular anxiousness and horn-blasting character of a bus-event. The point being, time, and space emerge as *attributes* out of these relations in much the same way as other bus attributes; time and space are all essentially formed in the same way and are made from the same stuff. Characteristics and qualities, such as noise and temporal franticness, must be viewed as *substantial*, and it is for this reason, Whitehead (1920) states that "time and space should be attributes of the substance" (p. 21).

Though we retain an enduring sense of character or sense of self that persists through series of events, like the ghostly object, this enduring character must be seen as an abstract generalization (Whitehead, 1920). I am, in reality, a series of events and a series of *self-at-the-moments* in the same way in which a bus is really a series of *bus-at-the-moments*, and who or what I am in traffic is not necessarily who I am when *not* in traffic. As a subject, I am the prehensions I experience from the settled past, whether they be from long ago or from the past few moments, and, whilst in traffic, these prehensions are quite possibly dominated by particular kinds of prehensions.

Yet, in contrast to this way of thinking, most sociological research continues to focus on the agency of the human subject as an independent, autonomous actor. Shove explains how most social research in the UK overemphasizes the cognitive autonomy of the human subject in its environment through a focus on the interior conscious world. She describes how this research focuses on changing people's habits by first persuading them to change their attitudes, viewing this as a direct causal factor towards bringing about change in concrete behaviour (Shove, Pantzar, & Watson, 2012). However, this approach fails to take into account *how* and *why* values and attitudes come to be what they are in the first place, a dilemma Shove calls the "value-action gap" (Shove, 2010, p. 1276). Such an approach views habits as causal drivers in themselves, rather than something that is allowed or disallowed due to the contingent sociotechnical configurations it is embedded in. Such approaches over-value conscious judgements, while undervaluing the vague, "primitive" (Whitehead, 1985, p. 43) presences experienced in the mode of *causal efficacy*, which are "heavy with the contact of the things gone by, which lay their grip on our immediate selves" (Whitehead, 1985, p. 44). In addition to this, when traffic users are engaged in absorbed coping, many actions may be more akin to *automaticities*, operating at the level of the unconscious, even while generating a feeling that we are in control of these actions (Wegner, 2005), perhaps unconsciously replicating, in a kind of empathy, the feelings from the immediate past.

The following story illustrates how 'the best of intentions' are formed subsequent to actual events and how the subject emerges prior to conscious narratives, as entangled in the complex flux of concrete actual events. Pham, a long-term veteran of the HCMC traffic, works as a motorcycle taxi driver, known in Vietnamese as a *xe ôm* driver, a job he has done for many years. He is in his 50s, exudes a calm, patient, and especially easy going demeanour and is quick to smile, an action that all but closes his eyes, leaving the narrowest of slits and revealing large uneven teeth. Viewing Pham's traffic video, I was impressed by how smoothly he moved through the traffic. From the GoPro footage, I could see that he always took his time and seemed to have a knack for avoiding dangerous situations, always seeming to be in harmony with the way things work.

However, Pham's occupation, at least in the format that he is familiar with, has all but disappeared. Ten years ago, *xe ôm* drivers could be seen waiting on almost every street corner, sitting on their motorbikes, perhaps smoking a cigarette, patiently waiting for a customer to hail for their services. These riders were generally based within one area and knew every *hẻm*, every alley, and every shop. When I first lived in HCMC, I used the same *xe ôm* rider every day to go to work, whom I suspect desperately needed eye-glasses and used all other senses apart from sight to navigate his way around. In those days, it was always common practice to stop and ask these icons of the streets for directions if one got lost; time and navigational advice were things they always seemed to have an excess of. In the intervening years, smart phones, Google maps, and ride-hailing applications such as Grab and Uber have completely changed the whole landscape, and spotting a *xe ôm* now is like spotting a diplodocus.

Since then, the streets have become a sea of Grab-green, and it seems the job of choice for young Vietnamese men is working as a motorcycle taxi and delivery rider, a job that offers flexibility and the opportunity to be one's own boss, without the shackles of an office desk. When the online ride-hailing model first began to disrupt the status quo in HCMC, Uber and Grab riders became the subject of beatings by gangs of traditional *xe ôm* riders, who saw the threat of being squeezed out of the market by new technologies (Tuoi Tre News, 2017). Uber and Grab employees, who would normally be considered each other's competition, were forced to band together in roadside groups in order to protect themselves from these beatings. Similarly, the major taxi companies also felt under threat and displayed stickers on their cars accusing Grab and Uber of unfair business practices. Though Grab has since bought out Uber in Vietnam, there are other new ride-hailing companies that have moved in, and riders from these different competing companies are still seen in unified collectives, parked on the side of the road, whilst taking a break and waiting for customers. The transparency offered in this new model, for example, of payment options and the ability to track riders, has also introduced new elements of safety and consistency that were absent in the earlier system, and this is probably one reason they have become so ubiquitous.

These *infrastructures of technicity*, such as Grab ride-hailing applications and Google maps, have completely changed the landscape of HCMC traffic in a few short years. Smart phones are now just another part of the traffic infrastructure, so much so, that it is a completely normalized practice to look into one's smart phone, whilst riding a motorcycle. Traffic videos taken by traffic users reveal an enormous number of riders and drivers in the traffic engaging with their phones. When I asked various traffic users to point out something in their traffic videos that had captured their attention, they almost never pointed out the motorcycle riders who rode with one hand on the motorcycle and the other holding a smart phone, with their eyes firmly glued to the screen (riders, especially professional motorcycle couriers, are increasingly installing smart phone holders attached to their handlebars, which leaves both hands mostly free, but still takes the eyes from the road). One exception, who made a point of speaking about smart phone users in traffic, was Ngân, the lottery ticket seller, whose job places her in constant danger of being hit by motorcyclists who are not looking where they are going. Even Duy, the bus driver, whose job is made very stressful by the more unpredictable movements of motorcycles, never commented on this practice of smart phone usage in his traffic videos.

Whilst watching the traffic video of Pham, the *xe ôm* rider, I observed a particular scene when he was turning right behind another motorcyclist ahead of him, and he beeped his horn in a situation I thought unnecessary. When I asked him about this, he told me he was using the horn to ask permission from the rider in front to allow him to turn first (it was, in fact, a short and rather polite-sounding beep). In fact, much of his traffic video revealed this kind of smooth, easy, and polite way of riding. Later in the video, we observed a particularly chaotic traffic scene at a large roundabout. Some of these HCMC roundabouts, especially if they have four or five busy roads leading into them, are like being caught up in a washing machine and then spat out the other side, not really knowing quite how you got there or where, exactly, 'there' is. In the video, we could see cars and motorcycles seemingly going in all directions at once, scrambling and scrabbling, in a 'me-first' kind of fashion. Whilst watching this scene, Pham explained that he felt annoyed that traffic users don't wait their turn and don't give way. He stated, "No one respects each other to stop to let others go in the traffic. . . . Everyone wants to go first. . . . It's really annoying." At this point, we paused the video (see Figure 8.3), and I asked Pham what he could do to make the situation better. He replied, "I would stop and wait and let everyone go first and then I go" and proceeded to point to vehicles on the screen and explain his plan: "I will stop and let these people go first. Let the taxi go and these two go and then I go."

However, despite these intentions and verbal propositions, when the video was played from that point, it showed that Pham did not wait for any of these other traffic users to go before him and, instead, signalled with his horn and rode in front of them, blocking their path with his bike so that he could go first. After watching this action, the research assistant, also present, reminded him in Vietnamese that he had not actually waited for the other riders, but instead, had ridden in front of them and not given way. She

Figure 8.3 Screenshot from video footage from Pham, the *xe ôm* rider, 2016.

then asked him to clarify as to whether he meant he would do it differently next time, or whether he was explaining what had happened this time; Pham then explained that his comments referred to how he would do it in the future, should the same kind of situation arise.

The kind of entangled absorbed coping that Pham was immersed in is very different to the atmosphere of the office where we watched his video. In the actual traffic, he was responding, generally unconsciously, to a range of affective influences, reacting unconsciously to the feeling-tones of the immediate past. In fact, as a relationally constituted entity, he was thoroughly immersed into these affective currents and structures. To a very large extent, we are our feelings of the experience of feeling the feelings that exist as data from a few moments ago. The conception of a true hybrid person/thing, a "humanized car," or an "automobilized person" (Katz, as cited in Thrift, 2008, p. 80) and the kinds of mediations that Latour (1999) describes as "a change in the very matter of expression" (p. 186) is possible only when both thing and person are constituted in ways that avoid the dichotomies of subject/object at the ontological level. Latour (1999) considers:

> The modern collective is the one in which the relations of humans and nonhumans are so intimate, the transactions so many, the mediations so convoluted, that there is no plausible sense in which artifact, corporate body, and subject can be distinguished. (p. 197)

For Latour (1999), the very notion of 'the social' is meaningless without recourse to objects and nonhumans (including objects such as bus timetables), and what we usually refer to as 'social relations' or 'social constructions' are, in fact, the systems of relations that form objects and human/nonhuman infrastructures through their different modes of existence, and it is these relations that give 'socialness' its durability over time, through relations so intimate as to be "almost promiscuous" (Latour, 1999, p. 204).

Noë (2012) suggests that the existential phenomenologists, particularly Heidegger and Dreyfus, set up, in opposition to each other, thought as presence, and 'unthought' as a kind of absence. Noë considers that the kind of 'unthought' awareness, characteristic of Heidegger's mode of *readiness-to-hand*, is both unthematized and unmediated, and that this stands, for Heidegger, "in sharp contrast to presence." Noë quotes Heidegger (1988), who says, "when we enter here through the door, we do not apprehend the seats as such, and the same holds for the doorknob" (as cited in Noë, 2012, p. 7). The door and doorknob are encountered as ready-to-hand because our aims lie in the room beyond them. If our aims were focused on the door, say in repairing it, it might be the hammer or screwdriver we are using to fix the door that would then be encountered in the mode of the ready-to-hand, thereby disappearing from our existential proximity.

However, whilst the door and doorknob may not have been consciously apprehended, they still exist in the world and are felt as prehensions and as series of events that have ingressed certain eternal objects. As with 'chair' (*chairness*), usually experienced accompanied with *sitability*, so too, doorknob is prehended as *turnability* or perhaps *grabbiness*, which is the nature of its prehensive form; as Nail (2019) says, "ontology is entangled with, and emerges from, its material conditions" (p. 12). However, over and above materialism, every prehension is, itself, a kind of mediation of which we are aware at a threshold beyond thought. Felt intensities are the primordial elements that influence the 'decisions' of subjects, and these feelings, according to Whitehead (1978), "*aim at* their subject," rather than *being* aimed at their subject (p. 222). The subject does not exist prior to any encounter with an external world, but emerges as feelings becoming subject, which is their aim. Rather than presupposing the existence of a subject, Whitehead's ontology "presupposes a datum which is met with feelings, and progressively attains the unity of a subject" (Sherburne, 1981, p. 16). In other words, the subject is both the feeler feeling *and* the feelings being felt, and the aim is always towards harmony, unity, and the integration of subject and world. It is through the process Whitehead calls the *phases of concrescence*, when new actual entities are born, that the openness of feelings from the settled facts of the past, come to be unified, and determined in a present becoming of a new subject, an outcome Whitehead (1978) calls the "satisfaction" (p. 26).

We are continuously immersed in this world of lures for feelings, of prehensions from the settled past, and subjective aims that compel us along particular trajectories. A car, a motorcycle, a chair, a hammer, a doorknob, and a sentence have their own particular prehensions, modes, forms, and aims, constituted of both physical and mental poles of feelings: carness, motorcycleness, sitability, swingability, and turnability. Just as the biological human body, is, itself, a "complex amplifier" (Whitehead, 1978, p. 119) or a translation tool (Harman, 2005; Merleau-Ponty, 2005), the amalgam of human with other-than-human functions in the same way. The nature of perception and awareness is *inheritance*, in the form of a feeling-tone, which retains the vector of its original character (Whitehead, 1978, p. 119): we feel and incorporate the feelings from the immediate settled past, and the body is how we place ourselves in the world, acting as our most primordial receptor of sensation. It is not a case of *cogito, ergo sum*, but rather of *sentio, ergo sum*, I feel, therefore I am.

Whitehead (1978) reminds us of a phenomenon that is so simple and obvious that it remains hidden: "we see *with* our eyes, we taste *with* our palates, we touch *with* our hands" (p. 170). The eye is a particular kind of prehension, and we have an awareness of this prehension in our experience of seeing; we feel *how* the eye sees, an experience that is so fundamental that it slips beneath conscious awareness, as it occurs before any objective intellectualization of the phenomenal experience of seeing. We are so immersed in our own bodies that we do not notice our constitutions by such prehensive forms. This intimate knowledge we have with our bodies, what Whitehead (1978) calls "withness of the body" (p. 81), is the starting point for our perception and awareness of the circumambient world, and it is also how we spatialize ourselves in the world. The causal efficacy or the feeling-tone of the eye as mediator, translator, and amplifier recesses into the background of awareness simply because we have become so intimate with the fact that our bodies are this way. This is non-sensory, non-cognitive, and non-conscious *perception*, and it is not as though I perceive and then infer the body's role in it (Winters, n.d.).

My body is the primordial *ground* for intellectual significance; it is the "fabric into which all objects are woven, and it is, at least in relation to the perceived world, the general instrument of my 'comprehension'" (Merleau-Ponty, 2005, p. 422). The senses are already unified without the need for a unifier (Merleau-Ponty, 2005), and so the nature of this experience is synaesthetic or, especially in the context of traffic, kinaesthetic. If the eye is a kind of prehension of which we are intimately intertwined and contributes to our constitution as a subject, so too are the prehensions that arise in conjunction with our intimate relations with nonhuman counterparts in traffic. As hybrid entities (driver/car, rider/motorcycle), the feeling-tones through which we navigate our traffic worlds are modulated and translated through prehensive subjective forms, shared with cars, motorbikes, and other causally efficacious modes. As our intimacy with our tools deepens, we feel *with* our cars, and we spatialize ourselves in the world through this hybrid form. Harman (2005) specifically describes how we *feel* whether the car will fit into the parking space, using the car as an extension of our bodies (p. 48). In the same way that we do not have a physiologist's knowledge of our own bodies, most people do not have a mechanic's knowledge of their cars or motorcycles. However, we *do* have a tacit working knowledge of both, including in an amalgamated or hybrid form; I feel with the prehensions I am absorbed in, including those that partly constitute my bike and car (and traffic lights, roundabouts, etc.), therefore, I am.

Merleau-Ponty (2005) describes how the hybrid intimacy with a blind man's stick is no longer an object for him and ceases to be perceived as such, as its point becomes "an area of sensitivity, extending the scope and active radius of touch, providing a parallel to sight" (p. 256). For someone who is blind, the prehensive qualities of the stick come to the fore, whilst the cognitive concept of the stick as an *object* is sublimated. To reiterate, the primordial ground from which emerge our perception, comprehension, and intellectual significance of the world is through intimate experiences with prehensions, not as a thinking thing in space, but as a subject that is actually constituted by those prehensions, including the nonhuman or inhuman aspects of our world.

The materiality of the thing, its particular subjective form, its enduring qualities, its shape, and its *handiness* also emerge in relation to other events, and, as Deleuze has written, "prehension is naturally open, open onto the world, without having to pass through a window" (Deleuze, 2006, p. 92). Rather than an ontological dichotomy of 'presence' and 'absence,' Whitehead's relationality of prehensions posits that what is discerned or discernible also includes the *negative* prehensions of *unactualized* potential – what would have been, but was not – which also serves to define the character of a thing. The point of the comparison of events is that awareness, including Heidegger's *circumspect* kind of awareness, enfolds presence and absence, and actuality and eternality. We feel the stress of a red traffic light due to the presence of *redness*, but, equally, we feel the stress of a *green* traffic light because of the lingering 'presence' of redness (it is green now, but it will soon be red). This phenomenon, what Massumi (2010) calls an 'affective fact' – in this case, a negative, not ingressed into an event – shows how events flow out from themselves in interdependent relations between factuality and eternality so that presence and absence can never be seen to be in a dichotomous 'either/or' relation or in opposition to the other.

References

Bennett, J. (2012). Systems and things: A response to Graham Harman and Timothy Morton. *New Literary History, 43*(2), 225–233.

Callon, M. (2001). Actor network theory. In N. J. Smelser & P. B. Baltes (Eds.), *International encyclopedia of the social & behavioral sciences* (pp. 62–66). Amsterdam, Netherlands: Elsevier.

Chabot, P. (2003). *The philosophy of Simondon: Between technology and individuation.* London, United Kingdom: Bloomsbury.

Dant, T. (2004). The driver-car. *Theory, Culture & Society, 21*(4–5), 61–79.

Deleuze, G. (2006). *The fold: Leibniz and the Baroque.* Minneapolis: University of Minnesota Press.

Goffey, A. (2008). Abstract experience. *Theory, Culture & Society, 25*(4), 15–30.

Harman, G. (2002). *Tool-being: Heidegger and the metaphysics of objects.* Chicago, IL: Open Court.

Harman, G. (2005). *Guerrilla metaphysics: Phenomenology and the carpentry of things.* Chicago, IL: Open Court.

Harman, G. (2007). Aesthetics as first philosophy: Levinas and the non-human. *Naked Punch, 9,* 21–30.

Heidegger, M. (1996). *Being and time: A translation of sein und zeit* (J. Stambaugh, Trans.). New York: State University of New York Press.

Hosinski, T. E. (1993). *Stubborn fact and creative advance: An introduction to the metaphysics of Alfred North Whitehead.* Lanham, MD: Rowman & Littlefield Publishers.

Latour, B. (1999). *Pandora's hope: Essays on the reality of science studies.* Cambridge, MA: Harvard University Press.

Latour, B. (2014). Another way to compose the common world. *HAU: Journal of Ethnographic Theory, 4*(1), 301–307.

Lefebvre, H. (2007). *Rhythmanalysis: Space, time and everyday life.* New York, NY: Continuum.

Manning, E. (2009). What if it didn't all begin and end with containment? Toward a leaky sense of self. *Body & Society, 15*(3), 33–45.

Massumi, B. (2010). The future birth of the affective fact: The political ontology of threat. In M. Gregg & G. J. Seigworth (Eds.), *The affect theory reader* (pp. 52–70). Durham, NC: Duke University Press.

Merleau-Ponty, M. (2005). *Phenomenology of perception.* New York, NY: Routledge.

Nail, T. (2019). *Being and motion.* Oxford, United Kingdom: Oxford University Press.

Noë, A. (2012). *Varieties of presence.* Cambridge, MA: Harvard University Press.

Shaviro, S. (2011). The universe of things. *Theory & Event, 14*(3), 1–13.

Sherburne, D. W. (Ed.). (1981). A key to Whitehead's process and reality. Chicago, IL: Chicago Press, Macmillan.

Shove, E. (2010). Beyond the ABC: Climate change policy and theories of social change. *Environment and Planning A, 42*(6), 1273–1285.

Shove, E., Pantzar, M., & Watson, M. (2012). *The dynamics of social practice: Everyday life and how it changes.* London, United Kingdom: Sage.

Thrift, N. (2008). *Non-representational theory: Space, politics, affect.* New York, NY: Routledge.

Tuoi Tre News. (2017, September 27). GrabBike driver attacked by 'xe om' group near Ho Chi Minh City bus station. *Tuoi Tre News.* Retrieved from https://tuoitrenews. vn/news/society/20170927/grabbike-driver-attacked-by-xe-om-group-near-ho-chi-minh-city-bus-station/41768.html

Urry, J. (2006). Inhabiting the car. *The Sociological Review, 54*(1_suppl), 17–31.

Venn, C. (2010). Individuation, relationality, affect: Rethinking the human in relation to the living. *Body & Society, 16*(1), 129–161.

Wegner, D. M. (2005). Who is the controller of controlled processes? In R. Hassin, J. S. Uleman, & J. A. Bargh (Eds.), *The new unconscious* (pp. 19–36). Oxford, United Kingdom: Oxford University Press.

Whitehead, A. N. (1920). *The concept of nature: Tarner lectures delivered in Trinity College,* November, 1919. Cambridge, United Kingdom: Cambridge University Press.

Whitehead, A. N. (1978). *Process and reality: An essay in cosmology (corrected edition)*. New York, NY: The Free Press. Originally published in 1929 by Macmillan.

Whitehead, A. N. (1985). *Symbolism, its meaning and effect*. New York, NY: Fordham University Press, The Macmillan Company.

Winters, K. (n.d.). A primer to Whiteheadian process thought. Retrieved from http://www.untiredwithloving.org/process_thought.pdf

9 Lures for driving and infrastructures of the ephemeral

Tết is the one major holiday in the Vietnamese calendar, and it is a time when almost everyone leaves the city for the provinces in order to be with their families. For two to three weeks before the holiday, the traffic in Ho Chi Minh City (HCMC) becomes, daily, more manic, propelled by plans and projects: holiday food to be bought and prepared, cleaning and housing renovations to be completed, and so on. Over time, a palpable atmosphere of affective urgency emerges in the traffic, passing from vehicle-to-vehicle, spreading and accumulating like a contagion, or rather, like an attractor, as riders, pedestrians, and drivers alike are drawn into its vortex of anxious logic. Then, just when you expect it to explode (just what 'it' would be, I am unsure), the air goes out of the whole thing like a deflating balloon and the city starts to empty out (relatively speaking), replaced by an eerie quiet that is as much presence as it is absence. Into these new vacancies of time and space, new layers of the city, once obscured by noise and the darting, flickering movements of traffic, begin to advance forward and catch my attention. For example, sounds: a leaf is blown across the road with a sound that seems, at first, so foreign that I don't recognize it. And sights: I find myself looking up above the street level – a dangerous activity on a normal busy day – and notice cosy-looking balconies and rooftop bars, looming large with a visual obviousness that makes me wonder why I never noticed them before.

The traffic in HCMC is *driven* (exponentially so), but driven by what? For one, atmosphere and affect, those fundamental and formative elements in urban traffic that receive such little attention in traffic research. Affect is a *complex* emergent outcome, but even traffic research that specifically aims at engaging with *complexity*, such as computer modelling, does so using tools that are often unable to capture or to account for affective atmospheres. The effect of geometrical concrete infrastructures and technologies is well studied, as are the effects of sounds, lighting, smells, etc. (Böhme, 2017), but the less tangible and more ephemeral infrastructures of complex emergent atmospheres, though they exist beyond individual senses and evaporate with reductionist analysis, are no less important in ordering and evolving a traffic system.

In her book, *The Transmission of Affect*, Brennan (2004) notes that the notion of the transference of affect was historically a commonly accepted fact, but this idea went out of favour as notions of a self-contained, biological individual began to dominate science. In *The Aesthetics of Atmospheres*, the German philosopher Böhme (2017) describes atmospheres as "quasi-objective or something existent intersubjectively" (p. 2). Böhme sees atmospheres as the glue that binds together subjects and objects, the result of the relationship between a feeling subject and material objective things, such as the geometry of a room, pictures, signs, sounds, and illumination. For Böhme, atmospheres can never be *things-in-themselves*, but are *mediators* between environments and our bodily feelings, always remaining "out there" (2017, p. 2). In other words, there is a kind of *in-betweenness* to atmospheres; they reside in relations (Gregg & Seigworth, 2010) or, better yet, *as* relations.

Atmospheres are not usually considered infrastructures, perhaps because of their lack of 'concreteness' and stability. However, Larkin (2013) says that infrastructures are "conceptually unruly" (p. 329) because they are both *things* and the relations *between* things. Larkin defines infrastructure as the "objects that create the grounds on which other objects operate" (p. 329). However, when thinking of infrastructure, the term 'relationalities' or perhaps *atmospheric infrastructures* affords more of an openness and conjures a less dualistic and more fluid arrangement of objects as infrastructural spatialities and temporalities that knit together both entities and environments. Atmospheres are synaesthetic and visceral, bridging thoughts, and things, denoting an immersion in the world that is both felt and thought, antecedent to emotion, but profoundly influential on emotion. In traffic, atmospheres are constituted by infrastructures and patterns of rhythms, forms, intensities, and resonances that make up events, and we emerge as subjects within events as amalgamations of these primordial ingredients.

Attendance to lures for feeling, embodied awarenesses, tacit skills, and attunements is especially relevant in HCMC traffic because it is a traffic system governed by a dual set of rules: those formally laid down by institutions, such as the government, and a perhaps, even more relevant set known as 'jungle' rules, the latter forming a system of tacit understandings that are fluid and always undergoing modifications. Due to the application of these informal or 'jungle' rules, negotiations in awarding right of way between vehicles in HCMC traffic events require a delicate (and oftentimes, not-so-delicate) balance of movement, energy, attunement, and intentionality, forming a shared set of tacit skills and knowledge unique to HCMC driving. The ability to anticipate the behaviour of others is based on our ability to know the states of minds of the people around us, and in HCMC traffic, this may be inferred by the most subtle of gestures, often mediated within nonhuman and inhuman relationalities, rather than, for example, a human facial gesture (especially given that almost all motorcycle riders in HCMC wear facemasks). For example, the use of acceleration into a space sends a message of intent (though when two vehicles both accelerate into a space simultaneously, in something of a game of 'chicken' – a regular occurrence in HCMC – both drivers need to be ready to pull back quickly). In myriads of other ways, drivers and riders, through their bodies and their vehicles, communicate with gestures such as nods, beeps, headlight light flashes, by physically blocking another, etc.

These informal rules have developed quite organically over time and continue to develop based on other changes in the traffic, such as in material heterogeneity and more ephemeral changes. As a dual system of governance, the formal and the informal rules are variously entangled, overlaid, and are also often misaligned with each other. Formal traffic laws can be unknown or unclear to traffic users, and traffic arrangements and signage systems are constantly being modified, so it is always better to err on the side of caution that the informal rule might prevail over the formal law, as the existence of two sets of rules creates many opportunities for misunderstandings and potential for accidents.

Phúc, the truck driver, said:

> In Vietnam, the traffic is really chaos and it is so complicated… people still follow the law, but traffic signals like markers and signposts are still limited so we just go" stressing that drivers must "use all the sense that we have.

Even professional drivers with many years of experience, such as bus drivers and truck drivers like Phúc, when asked about traffic laws, were often unclear on the official legal rules and often contradicted each other and sometimes even themselves when explaining them. Interestingly, most conversations I have had with HCMC traffic users concerning traffic laws usually result in a discussion on informal rules and tacit shared understandings, simply because the latter are more relevant for practical coping in traffic. For example, the ability to know which vehicle should give way in a given situation might be seen as fundamental to the operation of an urban traffic system, yet both the government bus driver and the truck driver I interviewed were unclear about priority vehicles in traffic. Duy, the bus driver, said,

> In Vietnam, there is no prioritized vehicle even if that is an ambulance with a loud sound on the street. That is about our awareness. If one is aware of this, he or she will give way.

Interestingly, the attitude and practice of *chen lấn*, a term mentioned earlier in this book, referring to a lack of self-awareness, was one of the most common expressions heard in interviews with Vietnamese traffic users; yet, according to Duy, it is awareness, rather than laws, that govern giving way, for example, to ambulances. This kind of flexibility is a fundamental characteristic in HCMC traffic. The implication from Duy is that it remains up to the personal decision of individual traffic users as to whether they give way to an ambulance or not. In fact, there *is* a law regarding priority vehicles (Vietnam Ministry of Justice, 2008), which can be easily accessed in English or Vietnamese online, and it clearly states that ambulances are priority vehicles when on duty.

A similar situation occurs in HCMC with 'zebra crossings,' those painted striped pedestrian crossings. Objects such as ambulances and zebra crossings might be viewed as universal and as stable forms that keep their qualities intact from place to place, but once we consider such objects in terms of being dynamic and fluid events, we better understand why the universal cannot be imported and mobilized as a 'free and clear' fact, with qualities and characteristics intact. Apparently, a hybrid offspring between a zebra and any other equine animal is called a *zebroid*, and this might be a more apt name for these painted crossings in HCMC. Pedestrians often choose to cross on a zebra crossing if there is one nearby, but doing so places one in a hybrid dual state of both safety and increased danger (see Figure 9.1). There exists a kind of aura of safety around these crossings that they somehow offer some – if only Platonic – protection, and yet drivers very rarely give way to pedestrians using them (though drivers might still consider a pedestrian on a zebra crossing differently to one who is not on a crossing, they just won't stop for them).

One scene in Nguyên, the taxi driver's traffic video, shows a pedestrian standing on a zebra crossing with his arms outstretched as the taxi drives through the crossing without giving way. When I asked Nguyên about this incident, he replied:

> I didn't [give way]. In my case, it is different. Even for some private cars, they also do not give way. Because for me, first, I have to go quick to save time so giving way for pedestrians is less likely to happen in our country. Everyone wants to go quick and if they see free space, they just go.

Figure 9.1 Pedestrians on a zebra crossing raise their arms in order to be seen by vehicles. Copyright 2016 by G. Wyatt.

Also present at this viewing session, Stephen, a motorcyclist originally from the UK, interpreted the gesture of the pedestrian as a sign of protest and annoyance that the taxi did not give way to him. However, this same gesture was understood by the taxi driver to mean that the pedestrian was motioning him to drive through. These two different readings are most likely due to the fact that the two users grew up in two different traffic systems, and there is an expectation on the part of Stephen that vehicles should give way to pedestrians when on such a marked crossing. Zebra crossings do not have innate qualities, subjective forms, or lures of feeling within them. Instead, each emerges out of its particular relationalities in its own unique way.

In another video, this time, from Duy, the government bus driver, a pedestrian was also seen standing on a marked pedestrian zebra crossing as the bus sped past her without giving way, leaving her stranded in the middle of the road (see Figure 9.2).

Again, this viewing session also included a non-Vietnamese traffic user, Laura, originally from Eastern Europe, who had only been living in HCMC for about two months. She obviously found this practice of not giving way to pedestrians on crossings strange, so she questioned Duy as to whether there was a rule about this and whether or not he knew of it.

DUY: In the past, when I learn how to drive, there was no rule like that. It was just the manner of giving way for pedestrians.
RESEARCH ASSISTANT (RA): How about now? Do we have a rule for that?
DUY: I don't know what is happening now. I don't have any updates. I attended the driving test a long time ago so I don't know when the government issues the rule so that I can update about it. You don't understand?

Figure 9.2 Pedestrians stranded on a zebra crossing wait for the bus to pass. Screenshot from video footage taken from government bus, 2016.

Interestingly, at this point, the atmosphere changed, and Duy, who had been extremely calm and friendly, now showed annoyance at being challenged on his knowledge of traffic law. It became apparent that Duy reserves the right to decide which rules he will choose to follow, as evidenced in the following:

DUY: In general, we drivers stop in some cases and go in other cases. It's about my attitude at that time. I don't give way for pedestrians all the time.

RA: What do you mean by your attitude at that time?

DUY: It's like if I want to stop, then I stop. Even just when that person gives me a nice impression, I would stop. So, it's not about stopping to let the pedestrians go first all the time. If I do this all the time, I will for sure be late for the journey.

These informal 'jungle rules' become enduring infrastructures, often in response to particular material arrangements, but, interestingly, when the physical materiality is modified or disappears, this *non-physical* infrastructural arrangement may still remain, as enduring patterns of practices or *lures for driving*. For example, the practice of vehicles turning out from smaller roads onto larger roads without giving way to other vehicles became common practice in the earlier motorcycle-dominated version of the system. Motorcycles have commonly done this, often without even looking for oncoming traffic (not to say that they do not use other senses), let alone giving way to it. However, as motorcycles transition to cars, the situation now exists that cars often replicate this practice of speeding out from side roads without giving way. Even more confusing, car drivers often straddle both positions, either signalling an intention to come out without giving way, perhaps through acceleration, but then suddenly stopping, or stopping half in and half out, or other ways of occupying both states. This occupying of all positions simultaneously might work well for the car driver, but it creates dangerous situations for motorcyclists who may ride into the side of a car that, without warning, comes out from a small street. Searle (1995, 2015) calls these kinds of non-physical enduring infrastructures "institutional facts." He writes that in order for an object to become an institutional fact, a kind of collective intentionality needs to be at work. Searle (1995) describes how a causally functioning *physical* wall may erode, but still function as a symbolic object or boundary marker, working, instead, in the form of a "status function" (p. 34), a barrier through which one could easily physically pass, but yet feels compelled not to due to the collective recognition of the status of the stones that create the boundary.

On one level, a zebra crossing exists as a matter-of-fact simple object, an object of a science laboratory, an object that should one come across it in one of those mini driving school environments constructed for drivers to practice, one would quite happily stop and wait for a group of imaginary school children to cross (perhaps led by an equally imaginary nun). However, concrete zebra crossings are not stable objects, but events, embedded in relations of subjective forms. The bus, as a physically mobile thing, embedded in relations that involve limitations on the available time to finish the route, has its own subjective forms, which involve an impetus that makes it very difficult for Duy to slow down for every little obstruction in his way. Whitehead (1978) describes the impetus of subjective forms in the following example:

> In our experience, we essentially arise out of our bodies which are the stubborn facts of the immediate relevant past. We are also carried on by our immediate past of personal experience; we finish a sentence *because* we have begun it. (p. 129)

Therefore, subjective forms have their own particular kinds of prompts, lures, or impetus. According to Merleau-Ponty (2005), 'seeing' is implicated into movement, not as objective transference in physical space, but as projects towards movement, as virtual movement. Movement begins with imagined ends that, at least virtually, are completed at the point where they began, just as the subjective form of a sentence begins with the projection of a completion point, an end goal, intrinsic in its aim, what Whitehead (1978) calls the "satisfaction" (p. 26).

In one video scene, we watch Laura riding her motorcycle towards a crowded intersection in readiness to turn right. Just before she entered the crowd of vehicles, I paused the video and asked her to recall from memory what happened next. She described how she 'felt' that the van in front of her had seen her, a feeling that guided her movements from then on. However, the video shows that it would have been very unlikely that the van driver would have been able to visually observe her, given her position in the traffic, and it seemed that this feeling was more like intuition. When recalling from memory the outcome of this event, Laura depicted herself as a kind of victor, riding through the intersection unimpeded, ahead of the van. However, once we resumed the video, it became apparent that this was not the case, and the video showed her having to stop as other vehicles turned in front of her and blocked her path, an outcome that clearly surprised her, judging by her response in the viewing session. As Laura entered the crowded intersection, her project towards movement had the end goal of a right-hand turn, and she was clearly surprised to see that the outcome of the event was different to how she remembered it.

Such subjective aims and lures for feeling are described as an "aesthetic impulse toward order, meaning and value" (Sherburne, 1986, p. 84), and as Sherburne (1986) states, each actual entity *is* this aesthetic impulse, both the feeler and the feelings being felt. In other words, subjective forms and lures are a mode of existence and expression that not only translate and mediate but also inspire aims and intent, significance, and purpose, and so we *feel* intentions and purposes in the same way that we feel emotions (Hosinski, 1993). In any beginning action or trajectory, we are already, in a sense, at the end. Subjective forms have their own kinds of impetus and carry us towards certain ends. Duy is, in a variety of affective ways, compelled *not* to stop for zebra crossings. There remains a tacit arrangement regarding zebra crossings in HCMC that drivers may stop if they feel like it or not stop if they don't, and the subjective form of the bus is to remain mobile, to finish what it has begun – to roll, to fulfil its essential being. Duy's job, which is already profoundly difficult, is made marginally easier in not giving way to people at zebra crossings, in not having to force a large moving object to slow down and stop for just one more obstacle. However, this belief in the power of universal abstractions persists, and tourists especially, are often left stranded on zebra crossings, with looks of disbelief, as the traffic swarms around them as though they are invisible.

Interestingly, the practical riding test to obtain a motorcycle rider's license in HCMC requires the user to engage with an object that is pure laboratory matter-of-fact, to ride through an environment that has been stripped of all the energy, atmospheres, logic, rhythms, and informal structures of the real traffic, and which requires a completely alternative skill set. This test does not require any of the kinds of engaged *attunements* necessary to negotiate the real traffic in HCMC. The test itself consists of riding slowly through a series of constructed artificial environments, including a painted pathway of two figure '8's, in which the rider is required to stay within the curved parallel lines, riding over some very low-level speed humps and navigating through painted zigzags. What is most curious of all in this laboratory of traffic fact/objects is why they chose to use these particular (and somewhat Platonic) shapes of figure '8's and zigzag forms, as they seem so irrelevant to the kinds of physical trajectories a motorcycle actually performs in the real traffic. This riding test is testament to what happens when a matter-of-concern object is reduced to a matter-of-fact object through reductionism and by removing all of the heterogeneity, mediations, enrolments, translations, and relations of the actual experience.

The thick present of a traffic event

Narrations offered post-event often attempt to break down and separate what Nail (2019) describes as the "thick" present (p. 372) and to offer discrete and linear steps to what was originally experienced as a continuous *durée*-like event. Viet, who is a professional van driver with 20 years of driving experience in HCMC traffic, narrated a video scene involving his vehicle and another car at a T-junction intersection. Whilst there are clear laws governing such situations, Viet only spoke of the shared collective intentionality at work: the jungle rules. He said, "in the driving profession, in such situations, we look at each other and can understand who would give way for who. . . . that's one skill."

Viet described the importance of certain cues and how this particular situation was resolved according to shared, informal gestures. He explained that if the car still proceeded and ignored these informal gestures and caused an accident, he would consider the other vehicle at fault. "That is the rule," he said, referring, not to the law, but to these shared informal ways of negotiation. In other words, the informal rules have real priority over the formal rules and, even for a professional driver, these informal gestures and cues are viewed as "official" and, in case of an accident, blame can be determined based on whether or not they were adhered to.

Viet said, "When he saw me flashing the headlight or heard the horn sound, he would give way for me. That is the rule. If he kept proceeding and a car accident occurred, he would be responsible for that."

Viet appeared to be an analytical driver and seemed to approach the traffic system as a giant puzzle to be solved. Therefore, in narrating the scene for me, he broke it down into linear steps. However, the event was probably experienced as a more continuous enfolding of all things into one thick present, instead of a discrete series of detached, independent abstractions. Rather, the appearance of the car approaching from the side road would have opened up a much more holistic and *circumspect* totality for Viet. In such a situation, we do not encounter this other entity in terms of objective and separate bits, such as a disembodied driver's face, independent vehicle lights or horn sounds, but rather as a much more kinaesthetic total experience.

When we walk into a room, as when we drive through a traffic environment, we do not put together the infrastructure of the furniture in the room piece-by-piece, but we go straight into it (Heidegger, 1996). Merleau-Ponty notes our ability to, for example, 'sense' the backs of objects even though they face away from us, or to 'see' objects from the perspectives of other objects, because we know how such relations work from past experience (Merleau-Ponty, 2005). In other words, the world is already there for us "in all things at hand" (Heidegger, 1996, p. 77). Space is, at once, a holistic, emergent existential totality and any encounter with an object brings into view this entire background/foreground of relationality (Heidegger, 1996). For example, when we see a motorcycle's turning light begin to flash ahead of us, it discloses and brings into view an entire world, as explained by Heidegger (1996):

> Signs are not things which stand in an indicating relationship to another thing but are useful things which explicitly bring a totality of useful things to circumspection so that the worldly character of what is at hand makes itself known at the same time. (p. 74)

This going straight into things with virtual intuitions and kinaesthetic sensibilities means that post-event narratives struggle to find words to explain just how we got to where we are or to verbalize the vague ways in which atmosphere affects and enfolds us when we are inside this thick fold of the event, as events are experienced through their essences rather than through facts. Sometimes, users explained events in their traffic videos using logic that was not necessarily part of the *being-in* of the event, as the conscious and the cognitive struggle to represent these more primordial awarenesses. However, fragments of such embodied awareness remain in verbal narrations, and many HCMC traffic users referred to transensory faculties that defied any single particular sense. Massumi (2008) describes these faculties as, "a direct and immediate self-referentiality of perception" (p. 6), which he calls "thinking-feeling," (p. 6) and which "slips behind the flow of potential action that the objectness suggests" (Massumi, 2008, p. 6). In a traffic environment like HCMC, users are always subconsciously reacting to aversive and adverse experiences, but post-event narrations are often constructed in ways that better align with our assumptions about who we think we are, in traffic or otherwise.

Stephen, the motorcyclist, originally from the UK, said the time spent riding his motorcycle "under the bright Saigon sun" on his way to work is the experience he most enjoys about his life here. However, whilst watching his own traffic video, he noticed a habit that no one else present at the viewing session noticed: he sometimes moves his thumb over to the turning light on/off switch located on the handlebars of his motorcycle (see Figure 9.3), not to turn the light off or on, he explained, but just out of habit. He said he had recently become aware of the fact that he regularly and mostly unconsciously checks the status of the turning light, as to whether it is in the 'off' position or the 'on' position.

Figure 9.3 As a bicyclist approaches going the wrong way, the motorbike rider's thumb moves across to touch the turning light switch. Screenshot from video footage, 2016.

He said, "I press it every, like, half an hour sometimes," then quickly corrected himself and said, "Not every half an hour. . . . ," eventually giving up on giving any kind of accurate estimation of how many times this thumb movement might occur. For the rest of the video viewing session, this practice involving the thumb was not discussed again.

Later, however, when viewing his traffic video alone, I noticed, once again, this thumb movement, and so decided – as actor-network theory would suggest – to 'follow the actor,' the actor, in this case, being an unsuspecting thumb. Upon closer examination, it was revealed that Stephen's thumb moved across to check the status of the turning light switch more than 68 times in a 35-minute motorcycle ride. I then edited the video, splicing together all of these 68 thumb movements into one continuous series, revealing an entire – no longer subtle – pattern of behaviour. This editing process revealed a sequence of embodied, felt transitions that were rhythmic in nature, and I could then see that the movements mostly occurred as Stephen was about to enter a tight atmosphere of traffic congestion or when exiting congestion. He also sometimes held his thumb on the switch throughout the entire congested scene, only to move it back when he was out the other side. In addition, this movement also occurred when another vehicle threatened to come into his path, no matter whether it was a large vehicle, such as a car or a bus, or a bicycle. Through this process, these minute movements could be seen to form a kind of rhythmic barometer for atmospheric change in traffic and offer a view of something that is 'invisible' because it is unstable, ephemeral, and not physical, namely the fluid infrastructures of atmospheres, patterns, and rhythms. The cumulative effects of such stressful affective atmospheres in traffic and in other domains of life would be exceedingly difficult to track or to reveal as linear cause-and-effect relations.

References

Böhme, G. (2017). *The Aesthetics of atmospheres: Ambiences, atmospheres and sensory experiences of spaces.* London, United Kingdom: Routledge.

Brennan, T. (2004). *The transmission of affect.* Ithaca, NY: Cornell University Press.

Gregg, M., & Seigworth, G. J. (Eds.). (2010). *The affect theory reader.* Durham, NC: Duke University Press.

Heidegger, M. (1996). *Being and time: A translation of sein und zeit* (J. Stambaugh, Trans.). New York: State University of New York Press.

Hosinski, T. E. (1993). *Stubborn fact and creative advance: An introduction to the metaphysics of Alfred North Whitehead.* Lanham, MD: Rowman & Littlefield Publishers.

Larkin, B. (2013). The politics and poetics of infrastructure. *Annual Review of Anthropology, 42,* 327–343.

Massumi, B. (2008). The Thinking-feeling of what happens. *Inflexions,* 1, 1–40. Retrieved from http://www.inflexions.org/n1_The-Thinking-Feeling-of-What-Happens-by-Brian-Massumi.pdf

Merleau-Ponty, M. (2005). *Phenomenology of perception*. New York, NY: Routledge.

Nail, T. (2019). *Being and motion*. Oxford, United Kingdom: Oxford University Press.

Searle, J. R. (1995). *The construction of social reality*. New York, NY: Simon and Schuster.

Searle, J. R. (2015). Status functions and institutional facts: Reply to Hindriks and Guala. *Journal of Institutional Economics, 11*(3), 507–514.

Sherburne, D. W. (1986). Decentering Whitehead. *Process Studies, 15*(2), 83–94.

Vietnam Ministry of Justice. (2008). Laws on road traffic, Article 22. Retrieved from http://www.moj.gov.vn/vbpq/en/lists/vn%20bn%20php%20lut/view_detail.aspx?itemid=10506#

Whitehead, A. N. (1978). *Process and reality: An essay in cosmology (corrected edition)*. New York, NY: The Free Press. Originally published in 1929 by Macmillan.

10 The emergence of order

What are the processes by which the infinitely complex flickering, rhythmic vibratory intensities of shimmering mobility become meaningful networks of abstracted simplifications imbued with significance, in other words, meaningful *order* out of pure randomness? Whitehead's definition of the notion of 'order' is a "society permissive of actualities with patterned intensity of feeling arising from adjusted contrasts" (Whitehead, 1978, p. 244). In other words, things bring like things. This statement suggests the possibility of autopoiesis, a circularity of – to use Maturana and Varela's original definition of autopoiesis – "networks of processes of production...of components" (as cited in Fuchs & Hofkirchner, 2010, p. 111) that produce other components that realize more of the networks of processes and relations that originally produced them and so on. Nothing comes from nowhere, and, in Whitehead's scheme, the 'something,' the limitation that comes from the all-possible everything, finds its directional impetus and form through either the prehensions of the past or the subjective aim towards the future, in ways that the system admits certain kinds of actualities and not others.

In order to put this into context, it is worthwhile to briefly summarize how a particular actuality comes into existence. Rosenthal (1996) says, "every actual occasion introduces novelty into the world by prehending and newly integrating what the past sends to it, and it does this via a 'lure' directed to the future through its prehension of eternal objects" (p. 544). Therefore, actual entities, in conjunction with eternal objects, have three aspects: (1) the datum or causes, which are the objectifications of the past; (2) the subjective character, which is the subjective form or aim, the lure – the 'how'; and (3) the concrescing (prehending) subject, which is, at once, the superjective character and the pragmatic value of its final satisfaction (Sherburne, 1981). Therefore, it is necessary to consider the nature of this 'lure' or subjective aim, and the criteria, processes, or forces that contribute to what prehensions or eternal objects are 'permitted' to go into the making of the new actual entity.

The processes by which the newly concrescing actual entity looks back towards the past are complicated, however, by the fact that the past is both open and settled and that the end is already innately in the beginning aim or form. As the concrescing actual entity looks forward towards the point of satisfaction or determination by way of the subjective aim, it is, in a sense, already at the end point when it begins, and, in looking back towards the objectifications of the settled past, though the past is fixed, it sees the past anew. The facts of the past are not truly 'settled' or fixed, until the completion of this process that results in a unified and complete subject, what Whitehead calls the *satisfaction*, and it is only at this point in the fulfilment of the aims, that the end result can be used to *retroactively* explain the account. In Stengers (2011) words, "the aspect of itself that will be taken into account, and thus contribute to the explanation of the account, did not, as such, preexist this taking-into-account" (pp. 186–187). Therefore, the aspects or characteristics of Event 1 (the objective facts in the past or prehensions) that are taken into account in the later event, Event 2 (the *concrescing* subject), did not exist before being taken up (*prehended*) by Event 2, because the aspects of the earlier event came to be viewed from a new standpoint or perspective. The past is 'fixed' as datum for a concrescing subject in the present, but the perspective of these objective facts of the past depend upon their *relation with* the newly prehending subject in the present. In other words, it is the nature of the relation that forms the substance of the world, and that particular relation, which has its own unique nature, is the bridge between past with future.

Everything is already ordered or unified, but ordered in multiplicities of potential, in fluid changing configurations of patterns and networks, and it is processes of experience that *unconceal* an already ordered universe. Order emerges in conjunction with what concerns us (Heidegger, 1996), and so we need to be meaningfully involved, absorbed into aims, goals, intentions, and purposes; as Alexander Pope's poem, Essay on Man, makes comment:

> All nature is but art, unknown to thee;
> All chance, direction, which thou canst not see;
> All discord, harmony, not understood. (*Essay on Man*, p. 36)

According to Whitehead (1978), "a society does not in any sense create the complex of eternal objects which constitutes its defining characteristic [but] only elicits that complex into importance for its members, and secures the reproduction of its membership" (p. 92). The key notion of the previous sentence is the 'elicits into importance.' The defining characteristics – logic, aesthetic, feeling-tones, and intensities – that hold together the nexus or society, *permit* or *elicit* certain kinds of ingressions (the nature of the relations act as attractors for similar kinds of relations), thereby inspiring concern and significance. The process of concrescence towards new actual entities is dominated by the subjective aim, which "essentially concerns" the actual entity as a fulfilled *superject* (Whitehead, 1978, p. 69), in a form that is "not primarily intellectual" (Whitehead, 1978, p. 85), but "is the lure or feeling"

(Whitehead, 1978, p. 85). In other words, purposes and intentions are, at the fundamental level, *felt*, rather than thought, and so are primordially entangled in subjective aims and lures of feeling.

As already mentioned, the unity, which is the 'satisfaction,' the outcome of the processes of concrescence, is already present in the subjective aim in its initial outset. Subjective aims have particular forms, such as the form of a sentence, the aim of which is the completion of the sentence or the lure of maintaining the impetus of a bus through a pedestrian crossing. So far, so good, however, Whitehead (1978) says, "the primary element in the 'lure for feeling' is the subject's prehension of the primordial nature of God" (p. 189), and this is where things get problematic. Whilst actual entities are *causa sui*, in that they are their own reasons and can autonomously 'decide' which datum to ingress or prehend and which to reject, they do so within the limitations inherent in the data provided from the past as well as via the mediating influence of God. For Whitehead, God *is* order, variously described as, "the foundation of order" and "the goad towards novelty" (Sherburne, 1981, p. 31), and it is he who arranges lures that guide subjective aims. Therefore, whilst Whitehead's God is not exactly the creator, he *does* arrange the conditions of possibility that stack the deck, so to speak, inspiring 'autonomous' creative processes towards new actualities. Therefore, the things that are temporal emerge by their participation with non-temporal entities: eternal objects – we should remember that a society does not create the eternal objects (Whitehead, 1978), as no new eternal objects ever arise (Root, 1953) – and this process is "mediated" (Hooper, 1942, p. 65) by an entity that combines temporal actuality with timeless potential, which is the primordial nature of God (Whitehead, 1978). Therefore, God acts as a kind of bridge between the value-less realm of the eternal and the actual realm of the meaningful, significant, and value-laden.

The question then becomes whether the Ho Chi Minh City (HCMC) traffic system, through its infrastructures of atmospheres and particular kinds of *lures for driving*, elicits and is permissive of certain kinds of ingressions and prehensions, thereby leading to certain kinds of outcomes, in and of itself, or whether a more central mediator that exists beyond the temporal realm, such as Whitehead's notion of God, might play a role as "the ultimate reason for all the conditions that make temporal actual entities possible" (Hosinski, 1993, p. 164) or the ground for limitation and, therefore, the "ground for concrete actuality" (Whitehead, 1948, p. 179). Following Sherburne, I suggest foregoing the whole notion of order as emanating from a 'centre,' such as Whitehead's God, but rather to conceive of a process whereby order, meaning, and value *coagulate*, 'clotting' like curds around increasingly dominant attractors, or, as Sherburne (1986) describes, spreading via "dynamic pockets or aggregates of order" (p. 84). Like spider webs, things attach to other things, building on each other, and, like memes, spread forms of style, aesthetic, and logic that hold together enduring rhythms and construct atmospheric infrastructures.

According to Whitehead (1978), the infinitely complex becomes meaningfully simplified, limited, and significant through a process he calls *transmutation* (p. 251). Through this process, any complex event becomes characterized or infused with overarching (or underlying) feeling-tones and affects that allow a multiplicity to be affectively experienced as a unity. Such forms and feelings, which provide definiteness and infuse events with character, are felt as much as thought, and so our experience of events is informed in ways that are synaesthetic, kinaesthetic, and initially pre-sensory. Transmutation is how an aggregate of entities forming a nexus come to be felt as one macrocosmic unity (Whitehead, 1978), for example, as a general feeling of *redness, motorcycleness, carness*, or *happy/sadness*. At this level of the primacy of affective intensities of feeling, paradoxes are accommodated because the experience is initially pre-cognitive and non-representational. These kinds of paradoxical feelings, such as happy/sadness, are difficult to label or verbally explain simply because they are not yet clearly corralled or represented as a 'this' or 'that' emotion, so a complex feeling of happiness and simultaneous sadness is possible even if difficult to put into words. This inadequacy to account for such paradoxes, according to Massumi (2002), is because there is no cultural–theoretical vocabulary specific to affect, and because affect operates according to different logics and different orders from that of emotion.

Transmutation is where order begins, and it is a process that necessitates a drastic simplification through the discarding of irrelevant detail, but, in doing so, brings about meaning and allows us to *feel* order in the world and to *feel* complex multiplistic intensities as a community or society (Whitehead, 1978), thereby merging into the conceptual. Transmutation is the process that allows affective intensities to flow across the discrete building blocks known as actual entities, infusing both subjects and objects with similar natures of being. This process of simplification into emergent subjective forms creates order as well as relevance in traffic, and such processes do not discriminate between the human, the inhuman, and the nonhuman. Therefore, when I am driving a car, I am entangled in an event that may be entirely characterized by *carness*, a subjective form that characterizes the driver as much as the car, and it is how our holistic, affective, and meaningful traffic relationalities are formed.

Phương – a university student who is quite new to living in HCMC – was asked, in a video-viewing session, to point out anything that was present in her traffic video that caught her attention, for whatever reason. She pointed out that all of the street food vendors were causing a disruption to the traffic by taking up road space and attracting customers to their stalls. When asked by the researcher to physically point to these particular food vendors in the video, she looked a little confused and, after some thought, said that the food vendors she referred to were not actually present in the video scene. Whilst there were many other factors causing traffic disruption in this scene, such as motorcycles – some of which were carrying large loads – and a large bus that had taken up much of the space on the street, none of these other factors were mentioned by Phương.

It turns out that in Phương's traffic topology, street food vendors contribute powerful negative influences, to the point where they do not even need to be physically or visually present (at least in the video) for blame to be attributed to them for traffic congestion. The 'presence' of these absent food vendors and their roles in traffic problems was a common theme for her in the video-viewing session. This may be because Phương had recently moved from the countryside to live in HCMC, and the complexity of the system had been transmuted in a way that simplified it for her, but also infused much of it with the negative influence of street vendors.

Later, another video scene showed Phương almost hit by a taxi that suddenly turned, without giving way, from a smaller side road onto the larger road that she was riding on. This scene was chosen by the researcher (the present author) and edited for viewing in its particular form in order to highlight the taxi, which was, for the researcher, the central actor in the scene. However, when asked to comment on this scene, Phương made no mention of the taxi. Instead, her focus was immediately on a black dog on the street, which she blamed for the series of events that led to the taxi almost hitting her (the dog was not even noticed by the present author until Phương highlighted it in the video). Through discussion with Phương, it emerged that the dog was 'guilty by association,' in that she had linked the troublesome dog with a physically proximate street food vendor. In this way, through processes of transmutation, the enormous complexity of this event was simplified, so that multiplicities came to be felt through an overarching feeling/concept of *streetvendorness*, a character that was also transmuted into the poor unsuspecting street dog.

We will never know whether the dog did, in fact, belong to the nearby street vendor, but the mobility of street vendors suggests it unlikely. Nevertheless, the dog emerged as significant, a matter of concern, not merely as any 'street dog,' but specifically a 'food-vendor-dog,' which meant that the dog came to be partly to blame for the incident. Interestingly, a much larger and seemingly more dangerous actor – the taxi – was never mentioned and receded into the background of Phương's awareness, in favour of a physically smaller, and from the researcher's perspective, seemingly more insignificant actor, a rather confused looking dog.

In order for the dog to become a matter of concern, Phương needed to be involved, in terms of intentions, purposes, aims, and goals; there needed to be something at stake. For this reason, Noë (2012) says that "the world shows up for us. But it doesn't show up for free" (p. 2). Perception emerges in conjunction with experience, which plays a central role in the development of understanding, meaning, knowledge and skills, and experience comes with, well, experience. Noë (2012) describes how a work of art, such as a painting may start out as a mystery, but as we learn more about the artist and the techniques, the context and the codes being employed, new horizons and levels of meaning open up to us. Like an artwork, the style of HCMC traffic both conceals and unconceals, and it is through becoming more familiar with its style that we more deeply and more meaningfully engage with this system of intelligibility and affect, whose emergent properties help to define the shapes, forms, and character of the entities within it, through being permissive of or eliciting certain kinds of ingressions.

Noë (2012) says, "we do things with style" (p. 3) and that questions of meaning and presence are really questions of style. Even more, it is not so much the case that we do things *with* style, but rather that we do things *through* style; the style is inseparable from what we do; it is the *how*. Style and aesthetic are the emergent qualities and characteristics of prehensions. Style is prehensions piled upon prehensions, and so style is fundamental to the ways in which we achieve access in our worlds through practical engagement, ways that also leave marks in the world that contribute and feed back into the style in a co-constitutive feedback loop. Like the enduring presence of arm movements in a Jackson Pollock painting, prehensions, modes, forms, rhythms, and the manoeuvres that constitute the HCMC traffic govern the way it organizes itself, and these things are also the concretization of cultural assumptions (Carlson, as cited in Edensor, 2010, p. 8). For this reason, driving a car is not something we merely do in an environment, but, as Stengers (2011) says, what we refer to when we use the phrase "driving a car" *is* the whole environment, and it is the mode and the style of being.

In District 1, in the middle of HCMC, just in front of the Bến Thành Market, there is a very busy roundabout intersection with many roads entering from different directions, and being next to a bus depot, it also has many buses traversing in and out. On one side of the roundabout, in front of the bus depot, there is a large open and unmarked area of blank asphalt-covered space. Across this space, traffic streams converge in diagonally opposing trajectories (opposing streams of traffic flow diagonally through each other in many locations in HCMC), and across these opposing streams, buses also drive in and out of the station (see Figure 10.1).

This large space is almost completely devoid of directional signage. Like a blank black canvas, it has no traffic lights or the usual navigational traffic markers such as lane markings or instructional road signs posted and seemingly few clear rules governing who should give way to whom.

Duy, the bus driver, traverses this area of traffic almost every day as part of his job and he knows it well. When this section of road appeared in his traffic video, I asked him if he thought the roundabout caused confusion to traffic users, given the absence of markers of any kind. He seemed a little confused by the premise of the question and replied that the area has many markers. Speaking through a translator, it was then made clear that I was referring to visual navigational aids, such as signage or markers, at which point, Duy agreed that there were, in fact, very few physical and visual navigational markers present in this area. Through conversation with Duy, it became apparent that whilst there were no marked dedicated lanes, he saw structure

Figure 10.1 Bus drives diagonally through opposing traffic flows without the aid of lane markings on the road. Screenshot from video footage taken from government bus, 2016.

and order where others might see nothing but chaos. However, the structure is somewhat fluid, and it arises from out of a quite simple tacitly accepted arrangement: each vehicle gives way to any other vehicle that moves into a space first, no matter what their trajectory or what kind of vehicle it is. Given that there is already a clear structure of informal or 'jungle rules' in place to cover this area, there is no real need for *additional* signage or structure. Such an arrangement can exist because users have a very limited range of options available to them: should another vehicle enter the path ahead of them, one needs to either give way or hit them and become involved in an accident, the latter being nobody's preference. Of course, this means that all space is up for grabs, and this results in hierarchical structures of priority that can be dependent on who has the most courage or who is in the bigger hurry.

Laura, also present in the session, likened the HCMC traffic system to Darwin's theory of evolution, following an order of power, beginning with trucks at the top of the list, followed, in decreasing order, by buses, taxi cars, cars, motorcycles, bicycles, and pedestrians at the bottom, who, she said, "have too many problems." When asked whether he thought the area needed more visual navigational aids, the bus driver responded, "They wouldn't know what marks to draw and no one would follow them anyway." He said that such visual aids might make the drivers drive "properly," but he noted that "properly" did not necessarily mean smoother or safer, and he considers that the safety aspect of this area is due to vehicles being able to "freely turn." With some pride, he then said that although he travels across this space every day as part of his job and has done so for many years, he has never witnessed an accident there. Travelling through this roundabout requires a high level of embodied kinaesthetic awareness, but this higher level of involvement also means that one is more engaged and more responsible for one's actions, more in the moment, so to speak.

In comparison, Stephen, the expatriate motorcyclist, described to me how he became frightened upon entering a roundabout whilst on holidays back home in the UK, because he felt he could no longer judge and predict the movements of the UK traffic users. Whilst the traffic in HCMC still appears messy to him, its flexibility and fluidity also involve, according to Stephen, a generosity of spirit, as it is a traffic culture with a willingness to forgive infractions of both the formal or informal order. Though the HCMC traffic might appear to newcomers as chaotic or random and unstructured, through experience, the various levels of order and the norms and expectations that prevent it from being true randomness become more apparent. Even with his experience, Stephen, when viewing the video footage taken from the front of a taxi, expressed almost disbelief at what he saw as chaos, whilst Nguyên, the Vietnamese taxi driver, when viewing the same scene, struggled to find anything out of the ordinary on which to comment.

In fact, the results of a comparative traffic study between three Southeast Asian countries found the traffic in HCMC to be more "disorderly" than the other two countries. Hsu, Sadullah, and Dao (2003) conducted a comparison study of the traffic culture of Vietnam, Malaysia, and Taiwan (Taiwan, especially, has a traffic mix similar to Vietnam, with a high percentage of the traffic being motorcycles), describing the differences in driving characteristics of the three countries. They noted a proliferation of 'disorderly' practices in HCMC traffic, such as motorcycles not respecting the stop line at traffic lights, making unsafe U-turns that cause conflicts with oncoming traffic, using horns incessantly, not keeping to motorcycle lanes, driving on the wrong side of the road and a lack of regulation to control lane-discipline or official enforcement to improve traffic order, and though this study is now quite old, the findings would still have relevance today.

However, traffic users have different tolerances and sensitivities for different kinds of disorder. Edensor cites Indian roads as an example where traffic is composed of variegated rhythms that compose a far less rhythmic pattern than would be familiar in traffic systems in the UK. However, 'disorder' should not be universalized as though it comes in only one 'flavour.' Though HCMC traffic appears chaotic and disorderly to some, especially visitors new to the city, the differently 'ordered-chaos' of the traffic in an Indian city might then seem chaotic to an HCMC resident.

The unique dynamics of the traffic in HCMC take some time to acclimatize to. Stephen commented to me, "it took the first six months of driving to forget everything that I knew to be right and do the opposite of everything that I know." Similarly, Laura, after riding a motorcycle in HCMC for two months, still finds the traffic unpredictable and confusing. In conversation with her, the source of this confusion appeared to be because she found the traffic system lacking more formalized signposting, such as directional signs and painted road markings. This lack of signage made her feel "heavy" in the traffic, she said, adding,

> "Everything that happens in HCMC is mostly unpredictable. You need to be careful. You need to have five eyes to look around . . . because it's unpredictable . . . absolutely unpredictable."

She expressed discomfort in being unable to "control the situation," a feeling that may stem from having to mould oneself into forces and dynamics that work to an unfamiliar or unknown logic, the artwork that remains undisclosed. However, Laura has a quick fix for this: she rides her motorcycle using a Google maps application on her smart phone, listening to the audio navigational directions through earphones. In doing so, however, she overlays a different kind of relationality over the existing experiential one, a topology of technicity that has been created by way of paradigms, such as what Whitehead calls 'simple location,' and thereby immerses herself in a world of prehensions and logics that are different from the experiential complexity outside of those parameters and relations of the Internet app.

Together with Duy, the bus driver, I watched a video of Laura navigating through the streets of HCMC on her motorcycle with the aid of the Google audio map. As we watched the video traffic sequences, Laura narrated her experiences moment-by-moment, and it became apparent that she was operating to rhythms and temporal structures that were dictated by the digital software, but that differed from the spatial and temporal world 'outside.' The Google application was constantly prompting her to turn a corner before she physically arrived and so was always a few steps ahead of herself, which changed the *style* of things and forced new kinds of attunements and awarenesses. Laura was simultaneously immersed in two very different modes of existence – topologies of technicity and topologies of experiential complexity, and this served to conceal some things and reveal others. The video showed Laura fearlessly speeding through situations, seemingly unaware of the dangers. Watching her traffic video, Duy, the Vietnamese bus driver, whose years in the traffic have given him considerable experience, pointed out many moments of danger for Laura and often stopped the video to ask if she was scared at that moment. Laura's lack of fear may have something to do with her limited experience in the traffic, or because the virtual topology of the Google app. was colouring her traffic world with a kind of safety net.

Traffic light entanglements

In HCMC, traffic light intersections are anything but simple events. Whilst watching a scene from the traffic video taken from the dashboard of Phúc's truck, we saw how various vehicles scrambled through the intersection in seeming defiance of the strict protocols by which the traffic lights operate. Watching this scene, Phúc said,

> At the intersection, the traffic light is not quite logical. For example, if people get stuck with obstacles, they have no time to wait as they are in the middle way and they have to move forward.

Of course, it is not the case that traffic lights are not 'logical,' but that their logic differs from the logic of the experiential complexity of the HCMC traffic, as the protocols of the traffic lights have emerged from more technical processes and paradigms. The kind of exactness and rigidity of a 'this' or a 'that' that they demand seems at odds with the more fluid and complex relationalities that constitute the being of the traffic in HCMC. The protocols by which the traffic lights are operating perform their own version of Whitehead's *transmutation*, simplifying and abstracting from the complexities of the traffic of which it has been delegated governance, demanding rigidity from entities that are constructed from more fluid, nonlinear interactions, and relations. Therefore, traffic lights do not allow for the kinds of complex situations in HCMC, whereby vehicles find themselves in less than ideal positions in the intersections. The paradox of imposing this exactness is that it often results in traffic users demanding flexibility and deviating from the more rigid structures through actions such as breaking traffic laws, for example, by driving through red lights.

For this reason, traffic lights can be a great source of anxiety in HCMC, a situation that is made worse, because most of the traffic lights have numerical countdown timers installed. These timers, in the form of an LED screen, display a countdown of numbers, so that drivers and riders can keep track of when the lights will change from red to green or green to red. This results in drivers and riders always counting down towards when the lights will change. What is interesting about these numerical timers is that they exist and influence real concrete events on two different levels, as the timers inhabit two different kinds of realities simultaneously. On one level, they count down numbers, which are abstract, universal Platonic objects, but in another sense, the timers also exist as what Massumi (2010) calls "affective facts." This means that the mere *presence* of the timers, regardless of the particular number being displayed (though the anxiety may get more intense as the numbers get closer to zero), infuses the whole intersection with a profound affective atmosphere of anxiety, though it is an affective dread that will always remain unactualized, always residing in the future. Massumi (2010) explains it thus: "fear is the anticipatory reality in the present of a threatening future. It is the felt reality of the nonexistent, loomingly present as the *affective fact* of the matter" (p. 54).

The timers come with their own kinds of prehensions, and the whole event can be ingressed by an eternal object and *transmuted* into a simple or complex feeling. If the countdown timers show that there will be a long wait (especially if the motorists are familiar with this intersection already), it compels the users to 'run' the red light rather than get stuck waiting for a long countdown sequence, or if already stopped and waiting, impatiently depart the intersection early before the light changes to green. This is a very common occurrence and creates a problem for those vehicles coming the other way, as they are forced to confront the oncoming traffic still in the middle of the intersection. On the other hand, if the traffic light countdown is short, this also creates anxiety due to the perception of the absence of time, causing the motorcycles to jostle for space in order to be first to depart the intersection when the light changes to green.

Anxiety that results from the perception of scarcity of time creates problems for HCMC traffic users on many different levels. For example, the water delivery motorcyclist, Huy, is seen constantly pulling his mobile phone out of his pocket in order to check the time when making a delivery. He commented that he has ongoing anxiety about time because there is an appointment time for the water bottles to be delivered, and if he does not arrive at that time, the client may leave the house, whereby he would have to return the bottles to the depot and set up another appointment. Huy said that anxiety related to time is the most challenging dimension to his job, a feeling echoed by Phúc, who said, "when I drive, I race with the time," and also by Duy, who said:

> When I ride a motorcycle, I can ride in any way that I want. But driving a bus is different. I have to drive very quickly like I am racing because I am afraid of being late. So, I have to rush. For riding a motorcycle, as I am such a type of person who rides a motorcycle very carefully, whenever I ride, I ride slowly.

Such feelings of anxiety are also collective emergences, whose affective presence is described in the following comment by Stephen:

> There is a massive urgency to get to where they need to get to . . . It's like everyone's got an appointment they're going to miss. That's the kind of feeling . . . and you get caught up in it.

One scene in Huy's traffic video (see Figure 10.2) shows traffic waiting at the traffic lights and reveals the extent of the influence of the traffic light countdown timers. Huy narrated the scene as follows:

> The traffic light here. . . the green light. . . If I recall correctly, it's just about more than ten seconds. It seems to have less running time so everyone wants to jostle to go quickly, to pass others and from the other road where the green is on so everyone also move forward so crashes often happen.

It is notable that most of the vehicles in this scene cannot even see what the actual numbers of the timer are, yet they still jostle for prime position,

Figure 10.2 Riders jostling for position at a traffic light intersection. Screenshot from video footage taken from Huy's motorbike, on route to deliver water, 2016.

stopping over the line in order to leave the traffic light intersection earlier than others.

Another example occurs in the traffic video of Thuy, who is a Vietnamese female motorbike rider in her 40s. The video footage shows her veering right out onto the wrong side of the road in order to negotiate a traffic light intersection. The road is small and in front of her are about 10 motorcycles waiting at the traffic lights, all of which she attempts to overtake. This is a dangerous move as there are cars and motorbikes coming in the opposite direction towards her, a danger that is further exacerbated by the fact that by her own confession, she is not a particularly proficient rider in terms of practical skills. When questioned about this movement, she said that she was worried that the traffic light was about to change to red. In fact, the light had just changed to green as she had arrived at the intersection, and the numbers on the countdown timer showed much more time than needed in order to pass easily through the intersection without the need to make a dangerous overtaking movement. Through the discussion that took place with her, it became apparent that the actual numbers on the traffic light timer were of no real consequence, but it was the mere presence of the timer itself that created anxiety for Thuy and forced her into a position that placed her in increased danger of physical harm.

References

Edensor, T. (2010). Introduction: Thinking about rhythm and space. In T. Edensor (Ed.), *Geographies of rhythm: Nature, place, mobilities and bodies* (pp. 1–18). Farnham, United Kingdom: Ashgate Publishing, Ltd.

Fuchs, C., & Hofkirchner, W. (2010). Autopoiesis and critical social systems theory. In R. Magalhães & R. Sanchez (Eds.), *Advanced series in management* (Vol. 6, pp. 111–129). Bingley, United Kingdom: Emerald Group Publishing Limited.

Heidegger, M. (1996). *Being and time: A translation of sein und zeit* (J. Stambaugh, Trans.). New York: State University of New York Press.

Hooper, S. E. (1942). Whitehead's philosophy: Eternal objects and God. *Philosophy, 17*(65), 47–68.

Hosinski, T. E. (1993). *Stubborn fact and creative advance: An introduction to the metaphysics of Alfred North Whitehead*. Lanham, MD: Rowman & Littlefield Publishers.

Hsu, T. P., Sadullah, A. F. M., & Dao, N. X. (2003). A comparative study on motorcycle traffic development of Taiwan, Malaysia, and Vietnam. *Journal of the Eastern Asia Society for Transportation Studies, 5*, 179–193.

Massumi, B. (2002). *Parables for the virtual: Movement, affect, sensation*. Durham, NC: Duke University Press.

Massumi, B. (2010). The future birth of the affective fact: The political ontology of threat. In M. Gregg & G. J. Seigworth (Eds.), *The affect theory reader* (pp. 52–70). Durham, NC: Duke University Press.

Noë, A. (2012). *Varieties of presence*. Cambridge, MA: Harvard University Press.

Pope, A. (1871). *Pope: Essay on man*. London, United Kingdom: Macmillan and Co.

Root, V. M. (1953). Eternal objects, attributes, and relations in Whitehead's philosophy. *Philosophy and Phenomenological Research, 14*(2), 196–204.

Rosenthal, S. B. (1996). Continuity, contingency, and time: The divergent intuitions of Whitehead and pragmatism. *Transactions of the Charles S. Peirce Society, 32*(4), 542–567.

Sherburne, D. W. (Ed.). (1981). A key to Whitehead's process and reality. Chicago, IL: Chicago Press, Macmillan.

Sherburne, D. W. (1986). Decentering Whitehead. *Process studies, 15*(2), 83–94.

Stengers, I. (2011). *Thinking with Whitehead: A free and wild creation of concepts*. Cambridge, MA: Harvard University Press.

Tuoi Tre News. (2016, May 9). Make traffic lights smarter to relieve Ho Chi Minh City congestion: Expat. *Tuoi Tre News.* Retrieved from https://tuoitrenews.vn/city-diary/34697/reduce-congestion-in-ho-chi-minh-city-make-traffic-lights-smarter-say-readers

Whitehead, A. N. (1948). *Science and the modern world*. New York, NY: The New American Library of World Literature.

Whitehead, A. N. (1978). *Process and reality: An essay in cosmology (corrected edition)*. New York, NY: The Free Press. Originally published in 1929 by Macmillan.

Epilogue

It was just over seven years ago when I first started riding a motorcycle in Ho Chi Minh City (HCMC). Knowing nothing about motorcycles or the riding of them, I bought a brand new red Honda Airblade, got a license – on someone else's bike – and more-or-less just hit the streets (I still have this same bike, which has probably now done about 60,000 HCMC-kilometres). Just for fun, I would sometimes wake early and go riding with the sunrise, staying out as the traffic slowly thickened around me with people on their way to work, and I became one with what seemed like the whole of humanity, flowing, *homogenously* mobilized. Sometimes I would ride and just let myself get lost and then try to pick my way back, anxious, nervous, but excited.

Then – though perhaps less so now – my experience in HCMC traffic was not so much one of 'mobilities' or 'commuting,' it was not scholarly, nor even an experience of the 'everyday.' Rather, it was simply an experience of coming face-to-face with humanity, with Vietnam, of diving into, and being carried along, as an ocean of humanity flowed in and out of its 'things,' its objects, embracing change along the way; to be in HCMC traffic on a motor-bike is to experience true immersion in something much larger than oneself. Riding in misty early-morning streets, filled with the smells of noodle soup being cooked in large pots on the sidewalk, past breakfast diners on small plastic chairs eating fried eggs and meat on sizzling dishes, all under the tall ancient trees of the HCMC streets, was, indeed, a privileged experience. It felt like home, yet strange and exotic, filled with hope and a sense of moving forward; every day was a new day, with a new, as yet unknown, destination.

I remember one day, still nervous on the bike, I was sitting deep in peak-hour traffic and waiting for the traffic lights to change, when I felt a tap on my back. I turned, thinking I was probably doing something wrong and was about to be told off, but everyone around me continued to stare straight ahead, oblivious to my enquiring looks. On the motorcycle next to me sat two Vietnamese men. As passenger on the back, was an old man, ridden, most likely by his son. This father and son team was close to me, certainly close enough for one of them to reach out and tap me on the back, but they showed no sign of even being aware of my existence. Just as I was about to give up my enquiries and resume my forward-facing demeanour, the old man gave me a sly look and smiled at me: it was a game. I smiled back and he reached out and rubbed my arm gently, as though to say, welcome, we are happy to have you here, enjoy your stay with us. It was such a simple gesture within such a complex environment, but it cut through, like a smile across a crowded room. And that was it; I was sold, destined to stay for the next seven years or more, destined to do my best to learn what everyone around me already seemed to know: the art of fluid resilience and the belief that all challenges, all problems, with time, will always work themselves out one way or the other.

Index

Printed in the United States
by Baker & Taylor Publisher Services